Femtosecond Laser Shaping

From Laboratory to Industry

Optical Sciences and Applications of Light

James C. Wyant, *Series Editor*
University of Arizona

Please visit our website www.crcpress.com for a full list of titles.

Femtosecond Laser Shaping

From Laboratory to Industry

Marcos Dantus

CRC Press
Taylor & Francis Group
Boca Raton London New York

CRC Press is an imprint of the
Taylor & Francis Group, an **informa** business

CRC Press
Taylor & Francis Group
6000 Broken Sound Parkway NW, Suite 300
Boca Raton, FL 33487-2742

First issued in paperback 2019

© 2018 by Taylor & Francis Group, LLC
CRC Press is an imprint of Taylor & Francis Group, an Informa business

No claim to original U.S. Government works

ISBN-13: 978-1-4987-6246-5 (hbk)
ISBN-13: 978-0-367-87744-6 (pbk)

Visit the Taylor & Francis Web site at
http://www.taylorandfrancis.com

and the CRC Press Web site at
http://www.crcpress.com

Contents

Acknowledgments

I am very grateful to Dr. Vadim V. Lozovoy who helped me prepare many of the illustrations.

COLLABORATORS

I take special joy in collaborating with a number of research groups and companies. I want to specially acknowledge the following collaborations:

Professors Sunney Xie (Harvard), Stephen Boppart (UIUC)
Dr. Jim Gord (Air Force Research Lab, CARS)
Dr. Sukesh Roy (Spectral Engines LLC, CARS, Machining)
Dr. Dmitry Pestov, Dr. Bingwei Xu (Biophotonic Solutions Inc)

FUNDING

I am especially grateful for funding received during my 24 years as a professor from the following agencies. Their funds have made it possible to have an active research group where we are educating the future generation and creating new knowledge: The National Science Foundation, Division of Chemistry (US), Chemical Sciences, Geosciences and Biosciences Division, Office of Basic Energy Sciences, Office of Science U.S. Department of Energy, DOE, The Army Research Office, The Air Force Office of Scientific Research, The National Institute of Health, The Department of Homeland Security, and the Michigan Economic Development Corporation.

Author

Marcos Dantus has pioneered the use of spectrally and temporally shaped ultrafast pulses as photonic reagents to probe molecular properties and control chemical reactions and for practical applications such as biomedical imaging, proteomics, and standoff detection of explosives. His contributions range from discovery of nonlinear optical properties and processes, invention of laser optimization instruments, and development of theory to simulate and predict the interaction of molecules with shaped laser beams. Dantus' development of an instrument capable of automated laser pulse compression is enabling research around the world as well as novel fiber laser designs. Dantus is a Fellow of the National Academy of Inventors, the American Physical Society, and the Optical Society of America.

Dantus received his BA and MA degrees in Chemistry from Brandeis University where he attended from 1982 to 1985. He received his PhD from Caltech in 1991 where he worked with A. H. Zewail on the development of Femtosecond Transition State Spectroscopy, now known as Femtochemistry (1999 Chemistry Nobel Prize to Zewail). He worked from 1991 to 1993 on the development of Ultrafast Electron Diffraction. He published 20 articles with Zewail during the period between 1986 and 1993. He has more than 200 publications and was named Inventor of the Year by Michigan State University given his 43 invention disclosures and 22 issued patents. He regularly collaborates with different branches of the Department of Defense and was invited to DARPA's Scientist Helping America, Arbitrary Waveform Generation, and Program for Ultrafast Laser Science workshops.

In 2012, Dantus was invited to speak on the future of Quantum Control for the NRC Committee of Atomic, Molecular, and Optical Science. Dantus has founded three companies: KTM Industries, a company that manufactures biodegradable packing materials. Biophotonic Solutions, the company that commercialized automated femtosecond pulse compression, where he served as chairman and Chief Technology Officer. More recently Dantus founded MTBIsense LLC, a company that is commercializing Rapid On-Site Sensing Headgear to prevent concussion related injuries in youth sports. Dantus is also the director of Research and Development of Total Power Inc., for which he formulated a biodegradable fuel additive used in the mining industry.

Prologue: Femto—A Universal Light Source

I was born soon after the laser was invented. Unfortunately, I had to wait until I was 16 to see one in action for the first time. That particular laser was being used to project a holographic image of the lunar lander. As many people have been, I was mesmerized by the shimmering quality of the light and the purity of the color. From the first moment I witnessed that laser, I knew that I had to find some career in the sciences that involved these marvelous instruments. Back in those days, lasers were only to be found in research laboratories; as they were expensive and very fragile.

Times have changed tremendously and lasers now play significant roles in a number of technologies. In fact, the Internet's ability to handle so much information is, to a large extent, dependent on lasers. Lasers are now used in manufacturing of parts ranging from tiny electronic components to welding cars.

As a sophomore in college, I sought the opportunity to work with lasers. I had a wonderful and generous undergraduate research advisor, I. Y. Chan, who took me under his wings, taught me about lasers, and gave me an independent research project, it was on the first high-resolution spectroscopy of fluoranthene, a polycyclic aromatic hydrocarbon found in tobacco smoke. I will not bore the readers with the ups and downs of a first research project. What is important is that I had my advisor's support and confidence to get me through the difficult steps. I am very proud of my first publication resulting from that effort.

When I arrived at Caltech, I met Professor Ahmed Zewail who had an exciting and ambitious dream in mind. He was convinced that one day lasers would be able to provide an answer to a fundamental question in chemistry: how long does it take to make or break a chemical bond? I was hooked, and Zewail was able to secure funding for the endeavor. I had the fortune to design the laboratory and purchase all of the equipment together with Dr. Mark Rosker. The laser, following the design of Charles Shank at Bell Laboratories, was built in record time. It took us only 1 month to go from an empty room to shooting femtosecond pulses, and we made the first breakthrough measurement in less than 5 months. Let me stress that, in those days, nothing came fully configured; in fact, we built seven dye circulators for the laser system from scratch. Pictures of the laser setup are shown in Figure 0.1. From that laboratory came a number of fundamental observations of chemical reactions. What is most important is that, for the first time, one could observe chemical bonds being formed or broken as if in slow motion. Zewail received in 1999 the Nobel Prize in Chemistry for his vision of probing chemical reactions in the fundamental time scale of atomic and molecular motion, namely, femtochemistry.

This book provides a fresh perspective on femtosecond lasers and their manipulation. Instead of just considering these lasers as short bursts of light and

FIGURE 0.1   Picture of the laser at Caltech ("Femtoland 1").

focusing on their brief duration, this book addresses other properties. In particular, this book highlights the bandwidth and wavelike properties of the pulses and how they interact with matter; this opens new scientific questions and technical opportunities.

The most important work related to this book came to me during a meditation on the possible applications of femtosecond lasers. It was the year 2000 and I was on the lookout for a new scientific direction. My motivation had to do with my desire to invent something that would be useful. That is when I had the significant realization that a short enough laser pulse could be used to produce light of any frequency—from microwaves to x-rays—a universal light source!

The motivation for such a light source was that one could use it to interrogate anything made of matter, from atoms to molecules, from solids to liquids and gases. What is even more exciting is that controlling such a light source would involve a programmable device that manipulates the waves making up the light pulse. This device then controls the time and spectral characteristics of the laser pulses, and can be used to control laser-matter interactions. The laser I imagined would require none of the conventional manual or mechanical optimization that can take a considerable amount of time. Therefore, one would have a programmable light source that would have universal utility for science as well as for never before possible applications. In terms of applications that could take advantage of the outstanding capabilities of this source, I considered cancer diagnosis, explosives detection, and proteomic analysis. I chose these three applications because I consider the saving of lives to be one of the most significant contributions one can make in their career.

The very fact I am writing this book is proof that the effort has been successful beyond my initial expectations. The planning for the research followed convention in the sense that I planned to start by following the same path of other scientists. However, I quickly changed my mind when trying to understand a "failed experiment" in our laboratory.

As was said by Louis Pasteur, "luck favors those that are prepared." The first experiment was designed to explore how different laser induced signals changed when the laser pulses were shaped systematically. The experiments involved a carefully designed set of shaped laser pulses and three different test materials. Interestingly, when I asked my students what the outcome was of the experiment, their first suggestion was that the experiment had not worked! When I asked to see the results, I found that signals were changing by a factor of three as a function of controlling the phase at certain wavelengths. This is when it all became clear to me and I knew that robust control of laser–matter interactions would soon become a reality. Based on that first experiment, the road to technology for perfectly controlled laser pulses and their applications toward cancer diagnosis, explosives detection, and proteomic analysis also became clear.

Imagine having the blueprint for technology to be developed during the next 20 years crystalize in your mind. It was hard to sleep because of the excitement. It was also very hard to have patience for setting the projects in motion. At that particular time, I had limited research funding; I did not have the necessary lasers or computer-controlled pulse shapers. Members of my research group preferred working on more conventional scientific projects than on dubious attempts to control chemical reactions with lasers. The skepticism was founded on claims from the 1970s, 1980s, and 1990s that such control had been achieved, but in the end, all of them failed. By the year 2000, it was essentially accepted that such control was so challenging that one was better off using a computer-based learning algorithm that started with randomly modified pulses and searched for the desired outcome. In those days, it was very difficult for me to communicate everything that I had figured out.

The bulk of this book outlines what I have learned about femtosecond lasers and laser–matter interactions. The book is written for those wanting to learn more about these lasers but find textbooks too technical and dry. For science and engineering students who want to see rigorous formulas, I suggest referring to the relevant research papers from my group. My goal is to make each chapter understandable to a broad audience. Most important for me is to convey the principle, and then to provide a perspective on how it can be applied to science and real-world applications.

The book is divided into the following sections. Chapters 1 and 2 introduce the laser in general and the femtosecond lasers in particular. Chapters 3 and 4 focus on what makes a laser pulse and discuss spectral phase in particular. Chapters 5 and 6 describe light–matter interactions. Chapters 7 and 8 focus back on femtosecond lasers, the generation of the short pulses, and pulse characterization. Chapters 9 and 10 introduce pulse shaping and how to use pulse shapers for pulse characterization. Chapters 11 through 15 introduce applications of femtosecond laser pulses that are already commercial or will soon be commercialized. Chapter 16 touches on some exciting new directions being contemplated, and Chapter 17 discusses some possible economic implications for the "femto revolution."

1 Introduction

It is likely that the majority of readers have already seen a laser. The availability of lasers as pointers for presentations and the use of lasers for light shows have made them quite common. This familiarity with lasers, however, is like our familiarity with microwave ovens; we know how they look and even how to turn them on but we do not understand how they work.

Lasers are very special sources of light. When we see light from a laser pointer strike a wall or a screen, we notice it is shimmering. Unlike a regular red or green light that smoothly varies from the highest intensity to the lowest intensity, laser light constantly shimmers from the highest intensity to the lowest, seemingly without smooth variation. The high degree of contrast is a direct result of the most important characteristics of laser light known by the term **coherence** (see Figure 1.1).

We start this part of the discussion by reminding the reader that light is more accurately known as electromagnetic radiation. One of the properties of this radiation is that it can be associated with fundamental particles known as **photons**. More importantly, for this discussion, light behaves as waves. When all the waves emanating from a source are perfectly in step, one says the source is coherent. The parameter that tells us if the wave is at its crest or at its trough is the **phase** (see Figure 1.1). Sunlight and the light from an incandescent light bulb are composed of waves that are not in step; one can imagine a random mixture of phases and colors.

One of the key properties of waves is that they **interfere**. For example, if two waves of the same energy and with the same phase are added, they interfere constructively. The resulting intensity, interestingly, is the sum of the two waves squared instead of just their sum. If, on the other hand, the two waves are out of phase, that is the crest of one coincides with the trough of the other, then the sum of the two equals zero intensity. This seemingly strange behavior of light leads to the shimmering property of laser scatter. The shimmering quality we observe is caused by **constructive or destructive interference** depending on the small variations in the surface being struck by the laser (see Figure 1.2). This is not observed when light that is not coherent shines on a surface. The random phases make it such that we only see an average scatter; it is highly unlikely to have a significant portion of the light achieving constructive or destructive interference.

Now that we have defined the laser as a source of coherent light, we shall briefly remind the reader how the laser was invented and what has motivated scientists and engineers to create short laser pulses. Given the fact that there are already numerous books that provide the history of lasers, I provide instead a pragmatic nonhistorical presentation. How to make a laser? First, in analogy to polarized sunglasses that allow only one polarization to be transmitted, one could

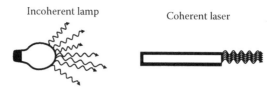

FIGURE 1.1   Coherent and incoherent light as well as phase.

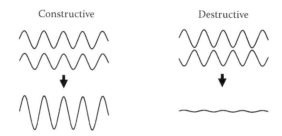

FIGURE 1.2   Constructive and destructive interference.

imagine a "coherence filter" that lets through only photons that are in phase. Interestingly, this simple approach has not been developed. Second, one could imagine a source of light that can be "triggered" such that light is emitted on cue, always in phase. This second approach is the one that worked.

The key to trigger coherent light emission was realized by Einstein and is known as stimulated emission. Stimulated emission can be thought to be the exact opposite of light absorption. However, when light is absorbed by some atom or molecule, emission is most likely to be spontaneous rather than stimulated. The likelihood that it is stimulated is greatly increased by the use of a cavity in which mirrors bounce the light back and forth multiple times in order to favor stimulated emission.

When the laser was first invented, it was seen more as a curiosity than a powerful tool with infinite applicability. Lasers were developed to test Einstein's theory. Many at the time said the laser was "a solution looking for a problem." However, it was soon realized that a laser projected a collimated beam of light to very long distances better than had ever been possible. This started a race to make the most powerful laser. Ideas of using lasers to cut through steel were very popular in cartoons and movies soon after the laser was invented, and decades earlier than they became practical. It took about 40 years for many of those dreams to become a reality. By the time I am writing this book, factories are now installing "second generation" lasers based on optical fiber designs that are much less expensive and rugged.

Laser intensity can be increased by compressing light into very powerful pulses. The concept of **peak intensity**, or the intensity of the pulse divided by its duration, is one that will concern much of this book. Imagine that a laser is capable of generating 100 W of light (Figure 1.3). If the light were restricted to a single pulse lasting 1 millisecond, then one would have a peak intensity of

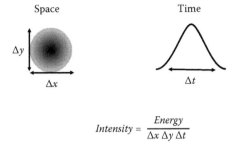

$$Intensity = \frac{Energy}{\Delta x\, \Delta y\, \Delta t}$$

FIGURE 1.3 Concept of peak intensity.

100,000 W. The laser can also be focused to a smaller area. Clearly, shorter pulses lead to higher peak intensities. Table 1.1, which shows pulse durations, will be instructive for the rest of the book. One can readily see that a 100-W 100-fs pulse would have a peak intensity of 10^{15} W, known as a petawatt. When such a laser source is focused, one is able to reach peak intensities exceeding 10^{21} W/cm^2.

A completely different motivation for the generation of short pulses came from the telecommunication industry. As scientists and engineers were contemplating the use of optical fibers and light pulses to communicate information, it became clear that the transmission length of a particular message depended on the duration of the pulses being used to communicate it. Making the pulses a thousand times shorter implied that one could transmit 1000 times more information. This was the motivation for scientists at Bell Laboratories to develop picosecond, and soon after femtosecond, pulsed laser sources.

As of the writing of this book, attosecond laser pulses have already been generated. While the intensity of these pulses is still very low, they are allowing unprecedented research into the dynamics of electrons. Similarly, a few daring scientists are already considering zeptosecond laser pulses capable of interacting directly with atomic nuclei.

TABLE 1.1
Times Units That are Much Smaller Than a Second and Their Names

Second	s	1
Millisecond	ms	10^{-3}
Microsecond	us	10^{-6}
Nanosecond	ns	10^{-9}
Picosecond	ps	10^{-12}
Femtosecond	fs	10^{-15}
Attosecond	as	10^{-18}
Zeptosecond	zs	10^{-21}
Yoctosecond	ys	10^{-24}

2 Why Femtosecond?

As reviewed in Chapter 1, there has been a constant drive to generate shorter laser pulses. In this chapter, we consider why femtosecond lasers led to the Chemistry Nobel Prize, a distinction that was not given when nanosecond or picosecond lasers were first used to study chemical reactions. We begin with our ability to time or clock physical phenomena. Some events take place in very short time scales and we need to have very fast clocks. For example, when we are watching Olympic competitions, we find that certain events are decided by hundredths of a second. The stopwatch used in Olympic events only has a one hundredth of a second resolution, and when two or more athletes arrive within that time, they are all awarded the medal corresponding to their time. There are many physical phenomena that take place in very short times, and this has inspired human ingenuity to devise faster and faster stopwatches.

Our ability to measure short-lived events using currently available electronics is still limited to the sub-nanosecond or picosecond time scales. To some extent, this is because the electrons being used by the electronic chips themselves to record the event need time to move from one transistor to the next. Some events in nature occur at time scales faster than nanoseconds. Measuring these events required the construction of a stopwatch using laser pulses. The operation of such a stopwatch involves the use of a first light pulse to initiate the process of measurement and a second light pulse to measure if the process being measured has been completed. This measurement process is also known as the pump-probe method; the pump pulse initiates the process and the probe pulse monitors the progress (Figure 2.1).

Measuring events with the pump-probe method takes advantage of the fact that the speed of light is a constant when transmitting in vacuum and that constant does not vary substantially in air. Deviations of the speed of light when it propagates in air and in other media will be discussed later in the book. Controlling the time of arrival of two pulses of light is achieved by carefully measuring the path that the laser pulses take to arrive at the object being measured. Given that the speed of light is approximately 0.3 m/ns, making small distance adjustments in the optical path of one of the beams controls the time of arrival between pump and probe pulses. With simple math, we realize that a 0.3-mm step corresponds to 1 ps and a 0.3-μm step corresponds to 1 fs.

The time resolution of pump-probe measurements is dependent on how small a step one can make with the optical delay line (Figure 2.2), and it is also dependent on the duration of the pulses of light being used. Lasers producing femtosecond laser pulses were first developed in the 1980s. This is approximately the time when I was visiting schools and deciding where to go to graduate school. In 1985, I was visiting the California Institute of Technology and had requested to speak

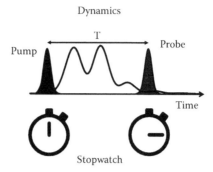

FIGURE 2.1   Concept of pump probe as a stopwatch.

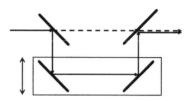

$$Delay\ time = \frac{Delayed\ path - non-delayed\ path}{Speed\ of\ light}$$

FIGURE 2.2   Concept of an optical delay line.

with Professor Ahmed Zewail. I had become acquainted with research from his group as an undergraduate. In particular, Zewail and his graduate student Peter Felker had just published a series of three papers studying the energy flow in a large organic molecule following picosecond laser excitation. Their findings were fascinating because they showed how energy oscillates between quantum states, a process known as quantum beats. Quantum beats had not been observed in such large molecules.

During my brief visit with Professor Zewail, he drew a molecule on a napkin and explained that when that molecule absorbs ultraviolet light, a specific chemical bond breaks. He then explained that while everyone knows that the bond breaks, no one knows how long it takes for a chemical bond to break. He explained that he was determined to find out. His enthusiasm was contagious and I told him that I would love to be involved in that project.

Turns out that the femtosecond laser provides the necessary time scale for measuring fundamental molecular processes such as the vibration of atoms in molecules and the time it takes to make or break chemical bonds. After a few interesting developments that I will not discuss here, I joined Zewail's research group in the summer of 1985 and worked on two different picosecond laser projects. Early in 1986, Zewail called me to his office and told me to design a femtosecond laser laboratory to carry out the measurement he had told me about during my first visit. This was a dream for many reasons. It was an opportunity to design a laser laboratory exactly the way I wanted. Having worked in a laser laboratory as an undergraduate and in another laboratory in Zewail's group, I wanted to make sure the new laboratory would be a clean room where you could easily find a small screw if it falls. By the end of that summer, Zewail received a grant to fund the equipment for the laboratory. He also hired Dr. Mark Rosker, who had already worked on one of the first femtosecond laser oscillators, to join our development. Mark and I built the laser based on designs from Bell Laboratories and were able to observe and follow with our laser pulses for the first time how a chemical bond breaks; something that had been considered impossible by many scientists.

Femtosecond laser pulses have had a very significant impact on how we now view chemical reactions. Bond breaking and bond formation are no longer considered instantaneous processes. Similarly, femtosecond pulses have also made significant impacts in related sciences such as biochemistry, biology, and physics. Expanding beyond chemistry, it is important to realize that femtosecond pulses have a duration that is faster than the associated speed at which atoms move in air or vibrate within molecules. Therefore, femtosecond lasers can interact with matter in ways that are very different from any other source of light.

Everything that is made of atoms, including all gases, liquids, and solids, has properties that depend on the speed at which atoms move. Given that atomic motion is closely related to thermal energy and that the femtosecond laser can act on matter faster than atoms can move, femtosecond lasers are thus able to evaporate atoms at a rate much faster than it takes for thermal energy to transmit through the object being vaporized. When a flame is used to cut steel, the entire steel object gets hot and melts. However, if a femtosecond laser is used to cut steel, the object remains cold. In fact, femtosecond pulses can cut without melting; therefore, they leave no scar of molten material. This ability of femtosecond lasers to vaporize atoms has led to what is known as non-thermal melting, a process that involves the use of picosecond and femtosecond lasers for cutting silicon wafers as well as other materials.

From a medical standpoint, an accidental exposure to an amplified femtosecond laser pulse led to the realization that these lasers can cut cleanly through the transparent cornea of the eye without causing collateral damage. This accident occurred at the University of Michigan, and the eye doctor that examined the scientist with the eye injury was sufficiently intrigued to initiate a conversation on how to use femtosecond lasers for performing eye surgeries. Material processing and corrective laser eye surgeries are presently the most important commercial uses of femtosecond lasers. These applications have already created multibillion dollar industries.

Scientific curiosity keeps taking advantage of femtosecond lasers, finding new applications and using these tools for solving new problems. In particular, part of the book will focus on how femtosecond lasers can be used for probing not only how atoms move but also how to control their motion. Laser control of matter can have many implications, from physical transformation such as the creation of nanometer features or nanoparticles, to chemical transformation, and finally to tailoring the light pulse so that it can best be used to detect cancer, for example. The U.S. Department of Energy's 5 Grand Challenges recognizes the importance of controlling matter and its processes.

As much shorter laser pulses are developed, beyond attosecond duration, it is almost certain that femtosecond lasers will be the engines that enable those new laser sources. This is why I consider femtosecond laser pulses to be critically important and they should be recognized as the first universal light source.

3 What Is a Light Pulse?

This book deals with short pulses of light; therefore, it is important to understand at a more basic level what a pulse of light is. If we consider a pulsed laser, say, one that produces pulses with duration of 100 fs, we can imagine each pulse to be a certain physical length given that we know the speed of light. In order to keep the math simple, we shall consider speed of light $c = 2.9979$ m/s to be ~0.3 μm/fs. Therefore, a 100-fs pulse is 30 μm long: roughly a third of the thickness of a human hair.

In this chapter, I want to discuss how a pulse of light has a finite dimension. To motivate the discussion, remember that the femtosecond pulse we are considering is made of light and that light is a wave. One interesting property of waves is that, in principle, they extend to infinite lengths. So how can it be that the light pulse is very short (in both time and space) when it is composed of waves that are very long in both time and space?

Answering the question will require us to remember the concepts of coherence, phase, and interference. The short pulse of light is formed by the constructive interference of light waves at the point of highest peak intensity; for practical and mathematical reasons, we will consider this to be the center of the pulse in space and time. Mathematically, we would consider the center to be the origin, as in the origin of the space and time coordinates. Thus, the coordinates of the peak portion of the pulse are zero in space and zero in time. The reason for this choice is that we can now imagine as an example a symmetric pulse decaying as a Gaussian or bell-shaped function in both space and time. Such a pulse is illustrated in Figure 3.1.

Destructive interference is the reason why the pulse intensity decays among its constituent waves. Note that if the pulse were made only of waves having a single wavelength, it would be impossible to arrange them such that they constructively interfere at some point in time and constructively interfere at a different point in time. However, if we have a collection of waves with different wavelengths, it is then much easier to imagine how it would now be possible to see constructive interference at the origin and destructive interference away from the peak of the pulse.

The previous discussion illustrates that the creation of a pulse of light necessitates light of different wavelengths that interfere constructively at one point and then destructively at other points in time. With some imagination, the reader can realize that the generation of *shorter* pulses requires a *broader* collection of different wavelengths. The breadth of wavelengths of a laser pulse is known as its bandwidth. The inverse relation between bandwidth and time has been known for a long time and is known as the time-bandwidth product (Figure 3.2). For Gaussian pulses, the pulse duration multiplied by the bandwidth expressed as a

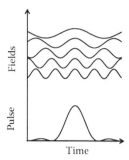

FIGURE 3.1   Concept of a light pulse.

Time-bandwidth product

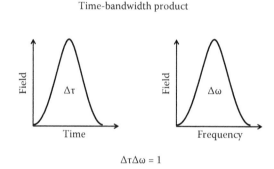

$$\Delta\tau\Delta\omega = 1$$

FIGURE 3.2   Concept of time bandwidth.

frequency is equal to 0.44. If we express the bandwidth in electron volts, then the time bandwidth for Gaussian pulses is approximately equal to ~2. Thus, a 1-fs pulse must have approximately 2 eV of bandwidth.

The previous paragraph is interesting because one could consider starting with a single-wavelength laser and using a that would block and unblock the beam very fast to attempt to generate a short laser pulse having all of its light having one specific wavelength. This would be impossible, and as we will see later, very fast chopping leads to the creation of new wavelengths of light. This observation implies that we have to consider that generating femtosecond pulses requires starting with a broad-bandwidth source of light. Presently, the best such sources are transparent media such as sapphire or glass doped with atoms that have optical transitions that are very sensitive to their local environment. This creates media that can emit light with a very broad bandwidth.

A transform-limited pulse occurs when all the wavelengths are coherent and interfere constructively at the peak of the pulse. Note that the shape of the spectrum is not what determines if the pulse is transform limited, it is only the phase that is important (Figure 3.3). As a first example, we can consider a 1-fs pulse with 1.9 eV of spectral bandwidth. Note that such a spectrum centered at 350 nm would encompass from the UV 200 nm to the near-IR 800-nm spectral range. All the different wavelengths of light within that 1.9-eV spectrum would need to be perfectly in phase at the pulse center. If such a pulse propagates in vacuum, then it remains short. If such a pulse propagates in air, however, it would quickly become much broader in time. The broadening would arise from the fact that the speed of light is a constant only for vacuum. In air or any other medium, the speed of light is not a constant. Variations in the speed of light owing to changes in the index of refraction cause dispersion, as we will discuss in detail later. For now, it is sufficient to know that we have to define the concept of spectral phase. The spectral phase function is an expression that tells us how the phase varies with wavelength. When all the waves in a pulse are in phase, we say its spectral phase is zero and the pulse is said to be transform-limited.

Mathematicians use different approaches to approximate a function: one of the most common ones is the Taylor or power series. In order to not alienate the readers that are more interested in getting the concept but are not planning to carry out their own calculations, I attempt to describe the usefulness of the power series to describe spectral phase, without using formulas. The power series is defined as a function of frequency, because frequency is proportional to energy, instead of wavelength. Frequency is simply 2π divided by wavelength. Readers interested in finding the formula should search for mathematical descriptions of spectral phase. Here, I describe this formula in words. We start by imagining any arbitrary function. The horizontal axis will correspond to the difference between the center frequency of the pulse and any frequency in the spectrum. This difference makes it easier to determine if we are talking about higher-energy or lower-energy photons.

In the power series, all elements depend on frequency except for the first, which is known as the absolute frequency of the pulse. This term, which corresponds to a vertical offset that anchors the function at the center frequency value, is important for very short pulses with fewer than three optical cycles such as those being contemplated in this chapter; therefore, I illustrate two pulses with different absolute phase in Figure 3.4. Notice how the magnitude and direction of the electric field changes between the three different pulses. In fact, the first pulse has a higher electric field than the second pulse and the field points upward. This is because the peak of the electric field coincides with the peak of the pulse envelope. The

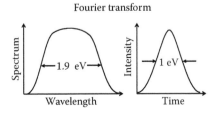

FIGURE 3.3 Spectrum and time profile of a transform-limited 1-fs laser pulse.

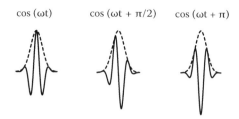

FIGURE 3.4   Concept of absolute phase.

reader can imagine that when a pulse is longer and has many more optical cycles, the absolute phase has much less of an effect when the pulse interacts with matter.

The second term in the expansion has a linear dependence on frequency. What is most interesting about the second term is that it corresponds to a time delay. To understand how this term is equivalent to delaying or advancing the pulse in time, I ask the reader to consider a pulse starting at a reference time that we will call time zero. The pulse we are imagining has a certain bandwidth and all the terms of its spectral phase are zero; it is a transform-limited pulse. Now, we need to imagine how such a pulse moves in space and time. We shall start with a single frequency and recall that light is a wave; therefore, every time the wave moves one wavelength (say 1.0 μm in length), the field oscillates up and down, as waves do, once. At this point, we imagine a second frequency within the bandwidth of the pulse. For example, this second frequency corresponds to a wave with 33% shorter wavelength than the first one (say, 0.66 μm). When this one moves one wavelength, it oscillates up and down once. However, when it moves 1 μm, the length of the first wave we first imagined, it oscillates one and a half times. Every time the pulse moves 1 μm, the first wave oscillates once and the second wave oscillates one and a half times. Clearly, the frequency of the second wave is 50% higher than that of the first wave. Most readers will agree that it is well known that the higher the frequency the more oscillations the wave makes; however, what few realize is that as a laser pulse moves in space, each of its frequency components oscillates at its own frequency. The number of oscillations that each wave executes depends linearly on their frequency. Thus, we can see that the linear spectral phase term corresponds to a time delay (when the sign is positive) and an advance in time when the sign is negative (Figure 3.5). Note that as the pulse moves in time, as long

A linear phase leads to a time delay

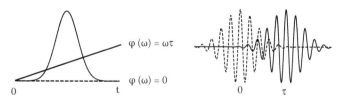

FIGURE 3.5   Concept of linear phase as time delay.

as it propagates in vacuum, it is not dispersed and remains as short as when it started. The more ambitious reader can search for software such as femtopulse_master to simulate the effect of specific spectral phases on a femtosecond laser pulse.

We now arrive at the third term of the expansion, the term that depends on the difference between the actual frequency of light and the average frequency of the pulse. For example, a short pulse has a broad spectrum. The weighted average frequency of the pulse, also known as the carrier frequency, is the reference point that allows us to speak about higher or lower frequencies in the spectrum of that pulse. The effect of a quadratic dependence can be explained by realizing that its parabolic shape (like the letter U) (see Figure 3.6) can be considered crudely to be composed of at least three lines. The first line, on the left-hand side affecting the lower-frequency components, has a negative slope. The negative slope implies those frequencies are being advanced with respect to the frequencies that are close to the central frequency of the pulse. The positive slope on the right-hand side corresponds to the higher frequencies which are delayed in time. The coefficient before the quadratic dependence indicates by how much the higher frequencies are delayed and the lower frequencies are advanced, hence by how much the pulse is stretched in time.

Interestingly, the second term of the expansion, in addition to being called dispersion, is also known as chirp. The reason for the name is that when the pulse has a large second-order spectral phase, the frequencies in the pulse disperse in time; therefore, the instantaneous spectrum of the pulse is changing in time. The average or carrier frequency of a positively chirped pulse sweeps from low to high; one can imagine a bird chirping. For a negatively chirped pulse, the carrier frequency sweeps from high to low.

The last term with which we need to concern ourselves for the moment is the one responsible for third-order dispersion, which depends on the cube of the frequency. The cubic function has the property that it will always have a positive or zero slope when it is positive, or negative slope when negative. That implies that all frequencies in the pulse, except for the carrier frequency, experience a delay or an advance. The magnitude of the advance or delay depends on the difference frequency cube. The cubic function and how the pulse looks when it has significant third-order dispersion are illustrated in Figure 3.7. Third-order dispersion causes a portion of the pulse to remain relatively intact while another portion,

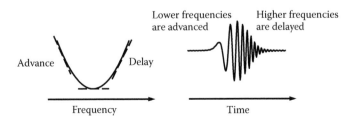

FIGURE 3.6 Concept of chirp.

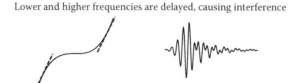

Lower and higher frequencies are delayed, causing interference

Frequency

Time

FIGURE 3.7 Concept of cubic phase or third order dispersion (TOD).

corresponding to those frequencies that differ most from the carrier frequency, sweeps ahead or behind the pulse itself. Furthermore, these frequencies cause an interference that is observed as oscillations in the electric field. Note that the oscillations are not periodic and that the electric field changes sign every time the envelope crosses a node.

A pulse has its shortest duration if all the dispersion terms are zero: that is, the phase across all the wavelengths is zero. Such a pulse is transform limited and provides the best time resolution. It should be noted that femtosecond lasers, despite all the engineering that goes into their design, rarely produce transform-limited pulses. Most femtosecond lasers come with a pair of prisms or gratings to minimize chirp (second-order dispersion). In Chapter 4, we will learn more about sources of dispersion.

4 When a Femtosecond Pulse Goes through Glass

The discussion about dispersion and spectral phase in Chapter 3 serves as a good academic introduction; however, it should not satisfy those who want to understand at a more empirical level how such changes in the spectral phase manifest themselves physically. Consider a 5-fs laser pulse with a carrier wavelength of 800 nm traveling a distance of 1 m in air. A measurement of such a pulse's duration would show it has stretched or broadened in time to 12 fs by transmitting in air. What has caused the pulse to more than double in duration? When light travels through a medium other than vacuum, each frequency component travels at the corresponding velocity of the said frequency in the medium. Because we are dealing with ultrashort pulses that have very broad bandwidth spectra, the difference in velocities experienced by all the frequencies causes the pulse to broaden (Figure 4.1).

The point above is sufficiently important that a second, more inclusive example is included. The fact that different frequencies of light travel at different speeds depending on the medium led to the introduction of a quantity known as the index of refraction. For example, the index of refraction of glass for 500-nm (green) light is 1.5. Therefore, when a light pulse with a 500-nm carrier wavelength travels in glass, it travels 50% slower than if it was traveling in vacuum (or to a close approximation in air). The delay of the pulse results from the second term in the Taylor expansion, the first term that depends on frequency.

Now, we return to the 5-fs pulse. Because of its broad bandwidth, we need to understand the functional form of the dependence of the index of refraction with respect to frequency. Such dependence is well known for optical media because optics manufacturers use it to create fine microscope objectives, telescopes, and camera lenses. A function describing the index of refraction as a function of wavelength, such as Sellmeier's equation, is then used to replicate the variations in the index as a function of wavelengths in one mathematical formula. The formula is needed because as the field travels in a medium that is not a vacuum, the spectral phase accumulated depends on variations in the index of refraction. If the index of refraction is a constant of frequency, then the pulse experiences only a delay proportional to the length of the medium. If the index had a linear dependence on frequency, the pulse would experience a delay and in addition, a linear chirp. Notice the progression: a constant index causes a delay while a slope in the index

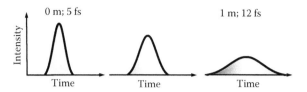

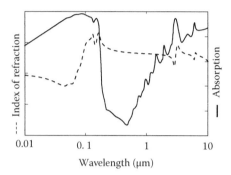

FIGURE 4.1 Concept of a 5-fs pulse broadened to 12 fs, after transmitting for a distance of 1 m in air.

FIGURE 4.2 Index of refraction and absorption of water (in logarithmic scale).

causes a chirp! Given the dependence of the index of refraction on absorption of light at certain frequencies, the relationship can be complicated, and therefore transmission in an optical medium can cause higher-order dispersion—third-order, fourth-order, and higher still.

The reader will appreciate that if one is dealing with relatively long 100-fs pulses at a wavelength far from an optical resonance in the medium, i.e., the medium is transparent, then it is enough to compensate for the delay and the linear chirp introduced by the medium, provided no higher-order dispersion was introduced by optical coatings or filters. Because this book is concerned with applications of ultrashort sub–10-fs pulses, one should consider the fact that the spectrum is broad enough to experience a linear chirp, third-order dispersion, and very likely higher-order dispersion. For example, we look at the index of refraction of water (Figure 4.2). The heavy line is proportional to absorption. We see that for visible wavelengths, around 0.5 µm, water is quite transparent. The dashed line in the same figure shows the logarithm of the index of refraction. Here, we see that in the visible wavelength, water has a relatively simple index dependence that leads to a delay, a chirp, and a small third-order phase dispersion. However, we see that in the UV and in the near IR, the index of refraction is quite irregular. The irregularities cause high-order phase dispersion on the pulse.

In addition to the changes in the index of refraction from the different optical materials, the use of special reflection and antireflection coatings that enhance the optical performance of lenses and mirrors can introduce unwanted additional dispersion. The reader can now understand why special optics designed for femtosecond laser pulses are often a good idea. These optics have a well-behaved index of refraction at the wavelength range of the laser spectrum. Another fact to consider is that the pulse spectrum for ultrashort pulses is seldom Gaussian. In many cases, the spectrum is quite irregular. In summary, the precise characterization of a laser pulse with a broad bandwidth spectrum after transmission through an optical setup can be challenging. Later in this book, we will discuss how one is able to characterize and correct imperfections in the spectrum and the phase of such pulses in order to deliver exactly the type of pulses needed for an experiment or an application.

5 Light–Matter Interactions (Part 1)

So far, we have devoted three chapters to describing light and femtosecond laser pulses, but we have yet to describe how light interacts with matter. We start our discussion here from the simplest cases and quickly build toward more interesting ones.

Consider a pulse of light propagating in air. As we discussed earlier, the air itself causes some dispersion that is approximately 1000 times less than that of glass for a pulse centered at 800 nm. What is happening in the air to cause such broadening? Given that light is an electromagnetic field, it interacts mostly with the electrons in the air molecules (primarily nitrogen and oxygen). Assuming the laser intensity is not high enough to disrupt the position of the electrons, but rather a minuscule perturbation, then light suffers a wavelength-dependent minor delay. The weak interaction between light and electrons is known as instantaneous polarization of the medium. As soon as the pulse moves to another region, the medium returns to its original state and no physical property of the medium is affected. For completeness, I mention that such instantaneous polarization leads to what is known as Rayleigh scattering. It is the wavelength dependence of Rayleigh scattering, greater scattering of bluer wavelengths than redder wavelengths, that is responsible for the sky looking blue during the day and red during the sunset. The wavelength dependence of the polarization of the medium causes pulse broadening as discussed in Chapter 4, and as illustrated in Figure 5.1.

Another closely related manifestation of instantaneous polarization is the reflection of light from metallic surfaces and even from the surface of well-polished glass. A metallic surface has loosely held electrons that are easily polarized by light. Once a pulse of light has been reflected, the electrons in a mirror return to their original state provided the light was relatively weak. Scattering from any surface is also closely related to reflection; however, we differentiate it from reflection because of multiple scattering surfaces. For example, dust particles reflect tiny portions of the beam in different directions and with different phases. In addition, there is a type of scattering process in which the resulting photons have a different energy from the photons in the laser pulse. The loss of energy results from a spontaneous quantum transition in which the medium gains energy, for example, through excitation of vibrations. This inelastic process is known as Raman scattering (Figure 5.2).

Often, the medium has certain degrees of freedom such as rotations, vibrations, and electronic transitions associated with its atomic constituents; the input laser field can excite those degrees of motion. In such cases, light may experience greater delays, and a portion of the light is absorbed. This phenomenon is greatly enhanced when the frequency of light coincides with a frequency

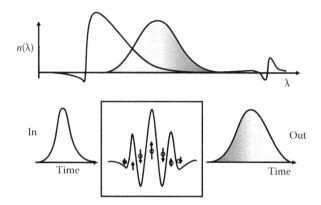

FIGURE 5.1 Concept of dispersion due to polarization.

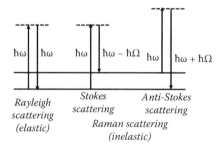

FIGURE 5.2 Concept of Rayleigh and Raman scattering.

associated with the energy required to excite a given quantum mechanical transition.

Femtosecond lasers, because of their broad bandwidth, can excite multiple transitions at the same time, thus creating coherent superpositions of quantum mechanical states. Unlike the "static" quantum mechanical states derived in conventional (energy resolved) quantum mechanics, wave packets have dynamic properties. For example, a vibrational wave packet, when plotted as a function of time, provides a fair representation of how a molecule vibrates. Similarly, when using polarized laser pulses, one is able to observe the rotational motion of molecules. The observation of vibrational and rotational motion in time (Figure 5.3) can be measured using pairs of pulses, as will be described below. Measuring such dynamic behavior of molecules with femtosecond laser pulses was one of the first important applications of femtosecond lasers to receive a Nobel Prize.

The ability of a femtosecond laser pulse to create wave packets, with a temporal resolution limited by the duration of the pulse, is what allowed the observation of a chemical bond being broken. A first femtosecond pump pulse can initiate atomic motion synchronously in the target molecules, and a second femtosecond

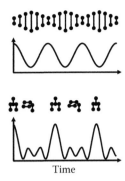

FIGURE 5.3   Concept of vibrational and rotational wave packets.

probe pulse can record the motion. For example, the excitation pulse can be in the ultraviolet range of the spectrum and the probe pulse can be in the visible range. In such an ideal case, the unexcited molecules can only absorb the ultraviolet pulse. Then, the excited wave packet can absorb the visible pulse. Each pump-probe time delay provides a snapshot of the wave packet motion. As the time delay between the pump and probe pulses is scanned, one obtains a movie of the molecular motion. The type of dynamics that can be recorded with this pump-probe approach depends on the wavelength of the laser pulses and on the time scale of the motions. Nowadays, femtosecond laser pulses can be prepared to have essentially any wavelength (as will be discussed in later chapters) and pulse durations have already reached the attosecond time scale.

So far, we have discussed linear laser–matter interactions. Quite simply, absorption, reflection, and scattering depend linearly on the intensity of the laser pulse. If the intensity of the pulse is doubled, the effect observed doubles as well. For example, if we shine twice as much light on a mirror, we observe twice as much light being reflected. This relationship, however, breaks down at high power densities and leads to nonlinear optical processes. One can think of this breakdown as an inability of the electrons to respond to light without distortion. When this occurs, the response of the medium is no longer linear, and we enter the much more interesting realm of nonlinear optics (NLO). Nonlinear optics opens a wide range of extremely useful possibilities as will be discussed later. One of the first NLO processes to be discovered was that intense light passing through certain crystals (those without a center of inversion) could generate photons with twice the frequency of the input light. The phenomenon of second harmonic generation (SHG) becomes very efficient when using femtosecond lasers, making it a very practical method for converting light from one color to another color, with half of the wavelength of the original.

SHG of femtosecond lasers should be thought of as sum-frequency generation (SFG), whereby all of the different frequencies in the laser pulse are summed with themselves to produce a pulse with an average frequency that is twice that of the

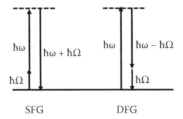

FIGURE 5.4 Concept of SFG and DFG.

initial pulse but with a broader bandwidth. As the reader can anticipate, there is a similar process known as difference frequency generation (DFG) in which one obtains the difference between the input frequencies (Figure 5.4). Both SFG and DFG can be carried out with one input beam, or with two different femtosecond laser pulses, or with one femtosecond pulse and one narrow-band picosecond or even nanosecond laser pulse. Finally, there are processes whereby the nonlinear optical medium can be adjusted to tune the wavelength of the output pulses.

There is a second class of nonlinear optical processes that are known as multi-photon transitions. When we discussed linear absorption, we considered the case where the frequency of light coincides with a quantum mechanical transition. Here, we consider the case where the frequency of light corresponds to half of the energy required to excite a molecule to an excited state. If two photons incident on the molecule are absorbed simultaneously, they can excite the molecule as if a single photon whose energy is the sum of the two previous photons had been absorbed. The possibility of multiphotonic transitions was first considered by Maria Goeppert-Mayer, decades before the laser was invented. The development of femtosecond laser pulses has facilitated the creation of pulses with sufficient peak intensity to easily excite multiphoton transitions. The introduction of multiphoton excitation to microscopy provided two significant advantages: first, it restricted the point of excitation to a narrow plane at which the intensity is high enough to cause excitation (Figure 5.5); second, it allowed the excitation of multiple different fluorescent molecules with a single laser. As the reader may have guessed, in addition to two-photon excitation, one may also induce three- and

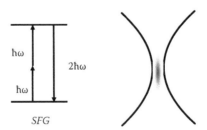

FIGURE 5.5 Concept of multiphoton excitation and MP microscopy.

four-photon excited fluorescence. Furthermore, there is a multiphoton transition that goes by the name of stimulated Raman scattering, in which the second pulse subtracts energy from the first in order to coherently excite a Raman transition. How one is able to manipulate a femtosecond laser pulse in order to control multiphoton excitation will be the subject of Chapter 11.

6 Light–Matter Interactions (Part 2)

In the previous discussion of nonlinear optics, we considered cases where one can envision the addition or subtraction of individual photon energies, e.g., as in two-photon excitation. Here, we consider the electric field of the laser pulse acting on the medium and exerting a force. A sufficiently intense laser pulse can modify the optical properties of the medium through its action on the electrons. When the field is weak, there is an instantaneous polarization that dissipates soon after the pulse (Figure 6.1). However, when the field is intense, it can cause significant displacement of the electrons and change the medium, at least during the time the laser is present. This implies that the laser can modify the medium strongly enough that the affected medium affects the pulse itself. These transient changes induced by the laser are known under the name "self-action" because the laser causes its own changes.

Of the self-action phenomena, the first to be discovered was the self-focusing phenomenon (Figure 6.2). A laser, as it propagates through a transparent medium, creates a change in the refractive index. Given that most laser beams are more intense in the center and their intensity trails off farther away from the center, the changes to the index of refraction are greater at the center and trail off radially. This spatial dependence implies that changes in the index induced by a laser pulse often resemble a lens (Figure 6.2). As the intense laser beam travels within a transparent medium, it focuses, and as it focuses, the power density increases and causes even more self-focusing. Within a short distance, the power density is sufficient to destroy the optical medium. Self-focusing can therefore have catastrophic effects. When intense pulsed lasers such as the YAG lasers were first developed, many an optic were destroyed because of self-focusing.

Just like self-focusing is a self-action in space, the instantaneous change in the index of refraction can cause a self-action in the time domain. Consider a pulse causing changes in the index of refraction that act on the rest of the pulse. When those changes occur on a time scale that is much shorter than the pulse duration itself, which is possible because the rise time of the pulse is by definition shorter than the pulse itself, then the pulse experiences spectral broadening. The relation between bandwidth and time was discussed in Chapter 3, in terms of the transform-limit. The change in the spectrum caused by self-action is known as self-phase modulation (SPM) (Figure 6.3). With femtosecond laser pulses, it is quite easy to start from a narrow bandwidth 10-nm pulse at 800 nm, which is essentially invisible to the naked eye, focus it on a transparent medium such as water, and to obtain a white-light output pulse that encompasses the entire visible spectrum from 400 to 800 nm. This effect is known as supercontinuum generation.

FIGURE 6.1   Concept of polarization.

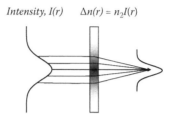

FIGURE 6.2   Concept of self-focusing.

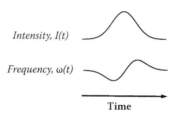

FIGURE 6.3   Concept of self-phase modulation.

Self-action phenomena can be very useful, especially for generating femtosecond laser pulses at different wavelengths and to convert longer pulses into shorter pulses. For example, an inexpensive compact laser producing relatively long 200-fs pulses can be compressed to sub–10-fs duration by first taking advantage of SPM and then by pulse compression. Self-action can also affect the spectral phase of the laser pulses. This is one reason why femtosecond lasers are very sensitive to changes in pulse energy. A reduction in the laser output of a few percent may change the spectrum and pulse duration in ways that are sometimes hard to predict unless all self-action phenomena are taken into account. Recent experiments have shown that self-action phenomena can act in unusually fortuitous ways. For example, the combination of focusing an intense femtosecond laser in air followed by its transmission through a glass window can lead to a four-times compression in pulse duration (Figure 6.4).

When the intensity of the laser pulse exceeds 10^{14} W/cm^2, the pulse ionizes most materials. The field is so intense that it pulls on electrons, usually managing

FIGURE 6.4   Concept of compression through continuum generation.

to extract the electrons that are held less tightly to atoms. When the molecules are isolated in a vacuum, single electron ionization is the first process that is observed. Increasing laser intensity or increasing pulse duration leads to the observation of multiple ionizations. It has been determined that at these moderate intensities, the electrons are pulled out of the molecule sequentially.

Beyond the simple picture given for ionization above, there is an interesting phenomenon. As the electric field pulls on the electron, the electromagnetic field of the pulse reverses polarity and then accelerates the electron toward the atom from where it originated. This phenomenon is known as re-scattering. The light emerging from re-scattering is easily identified because it has a clear energy signature equivalent to the number of photons of the incident laser that were absorbed by the electron during its interaction with the field. Values exceeding 50 photons were observed quite early; presently, it has been determined that the maximum values observed are for helium and that longer-wavelength lasers achieve greater acceleration. Light with kilo-electron volt energies, corresponding to about 1000-photon acceleration, has been reported. This process, known as high harmonic generation (HHG), is now used routinely in laboratories to generate x-ray laser pulses (Figure 6.5). Interestingly, the bandwidth generated during HHG is sufficient to support attosecond pulse durations. The field of attosecond science emerged around the year 2000 and has already yielded measurements of the coherent motion of electrons in molecules.

The above examples considered isolated small molecules or atoms interacting with intense light. A different scenario emerges when the molecules are in close proximity, such as in solids or liquids. In these cases, the front part of the pulse generates a few electrons. As the main part of the pulse interacts with the free electrons, it accelerates them and these electrons collide with the nearby

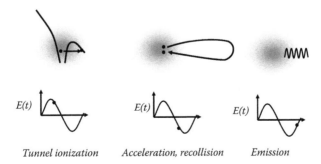

FIGURE 6.5   Three-step process for HHG.

atoms and molecules. The process generates more electrons, which can be further accelerated by the pulse. The process (generation, acceleration, high-energy electron-atom collisions releasing more electrons) leads to a process known as avalanche ionization, resulting in a large electron cloud moving very fast as a result of Coulomb repulsion and leaving behind a large number of positively charged atoms that also repel each other. The subsequent explosion is known as a Coulomb explosion, which leads to the ejection of the electrons and, soon after, the positively charged ions.

What is particularly special about avalanche ionization is that it can be confined to the surface of materials. This process can be used for ablating the top layer (few microns) without melting the sample (Figure 6.6). Because this process affects only the top layer of the material, it can be applied to absorbing, reflecting, and transparent materials. This process can also be accomplished within transparent materials, from microns to millimeters inside. In fact, this is the process that is presently used for femtosecond laser-based ophthalmic surgery. The femtosecond laser is able to cut the transparent cornea and shape it without burning the sensitive tissue before and after the focal plane of the laser.

When the intensity of the laser light exceeds 10^{18} W/cm^2, new phenomena are observed because the electron response becomes limited by relativity. One analogy that illustrates how one should think about the interaction of a pulse of light and a transparent medium is to consider how we are able to dip our hand into water (the linear optics regime). If we instead think of slapping the water surface with our hand as fast as possible (the relativistic optics regime), we find that water

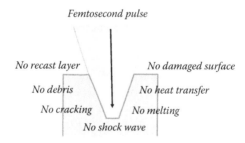

FIGURE 6.6   Concept of ablation.

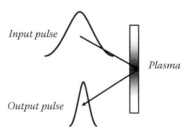

FIGURE 6.7   Concept of relativistic compression.

cannot conform to our hand fast enough and pushes back. The "compression" our hand experiences is analogous to the compression experienced by the laser (Figure 6.7). In fact, it has been estimated that a single cycle laser pulse with 2×10^{19} W/cm^2 would experience compression to 200 as (note that 1 as is one attosecond which is equal to 10^{-18} s).

Relativistic nonlinear optical phenomena are a subset of self-action phenomena and therefore share many similarities. Relativistic self-focusing has been observed, as well as self-compression. In terms of pulse compression, relativistic compression may be orders of magnitude more efficient than current methods for attosecond pulse generation based on electron re-collision. In fact, relativistic compression is considered the only viable approach to the production of zeptosecond (10^{-21} s) pulses, a goal that has yet to be achieved at the time this book was written.

7 About Femtosecond Pulse Generation

Having discussed many of the preliminaries, we are finally at the point where we can discuss femtosecond laser pulse generation. The simplest recipe for femtosecond pulse generation is to create a coherent bandwidth and to synchronize all the frequencies within the bandwidth so that their phases coincide at a given time. Sounds simple, but it is not.

For modest pulse durations, 100–500 fs, the bandwidth required can be simply derived from a number of luminescent materials such as semiconductors, metal-atom doped crystals and glasses, and laser dye solutions. The first femtosecond lasers producing ~100-fs pulses used a combination of a gain medium that generated the broadband light and a saturable absorber medium that was used to force the laser to produce short pulses. The balance of concentration, pumping power, and stability of the dye jets was very sensitive (Figure 0.1). Those early femtosecond lasers required constant adjustments and spontaneously stopped working every couple of hours. Within a decade, solid-state gain sources replaced liquid dyes. What is most surprising is that saturable absorbers were not needed. As a result, by the early 1990s, solid-state femtosecond lasers that could operate for tens of hours without intervention became available.

The generation of shorter pulses requires very broad bandwidths, exceeding 100 nm, and titanium-doped sapphire is the material that has proven capable of generating sufficient bandwidth to directly generate sub–5-fs pulses. This does not imply that all Ti:sapphire lasers can deliver such short pulses. For that, the laser cavity needs to be designed in such a way to keep very low losses within the entire bandwidth and to ensure all wavelengths experience the same amount of retardation. Given that sapphire itself causes uneven retardation, special coatings in the mirrors are needed to keep the pulse inside the cavity short. Special mirrors capable of reflecting light with different wavelengths at different depths have been introduced. Control over which wavelengths travel through how much material leads to tailored dispersion of femtosecond laser pulses. These so called "chirped" mirrors have simplified the design of femtosecond laser oscillators because prisms are no longer needed (Figure 7.1).

Having generated the bandwidth capable of supporting sub–10-fs pulses, it is important that all the frequencies within the laser are synchronized. In Chapter 4, we discussed how a femtosecond laser pulse is broadened by dispersion because the index of refraction is wavelength dependent. It is thus necessary to control the dispersion experienced by the laser pulses. It is quite remarkable that Treacy, in 1969, set out to solve the problem of dispersion control for femtosecond laser

Colliding pulses dye laser with saturable
absorber and prism compressor

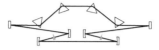

Ti:Sapphire with chirped mirrors

FIGURE 7.1   Evolution of femtosecond oscillators.

pulses. He did it at least one decade before lasers capable of producing sufficiently short pulses to require his solution were developed. In the introduction to his articles, he wrote that it was a purely academic pursuit, but one that he was sure would be useful. Treacy showed that a combination of prisms or gratings could be used to introduce a variable amount of positive or negative dispersion (Figure 7.2). Treacy's prism arrangement has played a role in the majority of femtosecond lasers.

When intense femtosecond pulses are required, it is possible to amplify them by passing the pulse through a gain medium. Each pass through the gain medium results in a 10-times increase in energy. There are many creative amplifier designs, some of them keeping the use of dispersive media to a minimum, and some with large amounts of dispersion that need to be compressed later (Figure 7.3). Typical amplifiers result in million-fold gain in the energy per pulse. There are amplifiers

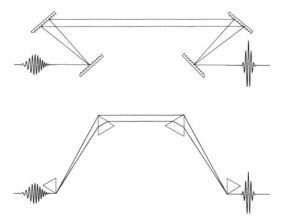

FIGURE 7.2   Prism and grating compressors from Treacy.

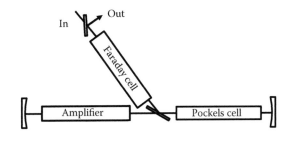

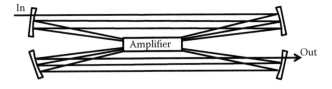

FIGURE 7.3   A couple of amplifier designs.

capable of taking a nanojoule pulse and output pulses with several joules of energy. Note that, in most cases, the femtosecond laser pulses produced in an oscillator at a repetition of 100 MHz are amplified and the repetition rate is reduced to a kilohertz or less.

Given that pulse energies increase by orders of magnitude within an amplifier, the pulses rapidly reach peak intensities capable of inducing self-action processes as discussed in Chapter 6. Recalling the discussion on self-action effects taking place in a medium, the reader would be perceptive to realize that femtosecond amplifier designs need to mitigate these processes in order to preserve the coherence of the pulses and to prevent destruction of the optics within the laser. The most successful approach to mitigating self-action processes to date is known as chirped-pulse amplification (CPA). A CPA amplifier uses a pair of gratings to stretch the pulse by at least a factor of 1000, through the addition of chirp. The pulses are then amplified, without fear of causing internal damage to the laser optics, and then compressed by a grating pair compressor, such as the one shown in Figure 7.2.

The amplification process leads to what is known as gain narrowing. Basically, the gain curve is limited by the amplification medium. Therefore, some spectral regions of the pulse experience greater amplification than others. Typically, the central wavelengths match the gain medium spectrum, leading to an amplified pulse spectrum that is more intense in the central wavelengths but less intense, and thus narrower, in the wings. Even for Ti:sapphire, gain narrowing limits pulse durations to about 40 fs. Fortunately, supercontinuum generation induced by 50- to 300-fs pulses in hollow waveguides and photonic crystal fibers has been shown to generate ultra-broad bandwidths that support sub–5-fs pulse durations.

More recently, amplification has been accomplished by using a nonlinear optical crystal that is pumped by an intense long-duration pulse. The chirped input

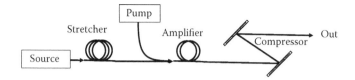

FIGURE 7.4   Fiber-CPA laser setup.

pulse is then amplified through a parametric process that can support very broad bandwidths. These amplifiers are known under the name Optical Parametric Chirped Pulse Amplifiers.

There is one important class of femtosecond lasers that has not been introduced yet. Fiber lasers, those in which the pulse generation and amplification take place within optical fibers, have quickly become standard in industrial and biomedical applications. The principal challenge for creating a femtosecond fiber laser was mitigating dispersion introduced by the fiber, and maintaining the power density below the level at which self-action processes destroy the pulse or the fiber itself. The concept of fiber-CPA, in analogy to CPA for the conventional lasers, has allowed the generation and amplification of femtosecond fiber laser pulses to millijoule levels (Figure 7.4). Often, the final step consists of a Treacy type of grating compressor.

In summary, femtosecond pulse generation requires an oscillator capable of creating highly stable seed laser pulses. Very compact fiber laser oscillators have replaced the cumbersome dye lasers of the 1980s for this task. The seed pulses then require stages of amplification and bandwidth generation (if pulses shorter than 200 fs are desired). Regenerative amplifiers and optical parametric amplifiers, taking advantage of CPA, have by now become the norm. Highly efficient femtosecond lasers presently take advantage of ytterbium-doped optical fibers. The final stage of most lasers is a grating compressor.

The dispersion control methods discussed so far are all passive. They are arrangements of optics that are placed in the laser system to address a certain amount of dispersion. Passive phase control has been used to deliver very short pulses; however, the pulses have some remaining high-order dispersion that cannot be corrected even with combinations of prisms and gratings. In Chapter 9, we will discuss adaptive pulse shaping; technology to control any type of phase irregularities affecting the laser pulses. Before we reach that chapter, we first need to discuss how to measure the pulses themselves.

8 How to Measure Femtosecond Pulses (Part 1)

Femtosecond laser pulses are much harder to measure than picosecond or nanosecond pulses. Nanosecond pulses can be easily measured using a fast photodiode and an oscilloscope. The fast photodetector produces a pulse of current that mimics the optical pulse, and an oscilloscope capable of reproducing the pulse traces it on a screen. Such technology is presently limited to ~10 ps and is not compatible for measuring sub–10-fs pulses.

There is a simple approach used for pulse measurement. The concept is based on the interferometer, a device used in 1887 by Michelson and Moreley. The idea is that light is sent along two different paths and then recombined. When both paths have the same length, interference between the two paths is observed. Michelson and Moreley wanted to test if the rotational and translational speed of planet Earth affected the speed of light. They found it did not, which implied the speed of light is always constant. This finding led Einstein, after considerable thought, to arrive at the theory of relativity, but I digress.

The interferometer was used to measure picosecond and, as soon as they were first generated, femtosecond laser pulses. The laser pulses are split into the two arms of the interferometer using a partially reflecting mirror. The pulses are later recombined, and one of the arms of the interferometer can be adjusted in length (Figure 8.1). When the length of that arm is scanned, one is able to see the interference between the pulse on the fixed arm and the one on the adjustable arm. It is interesting to note that such "linear" interferometric measurement cannot provide an accurate measurement of the laser pulse. In fact, if instead of femtosecond pulses one used incoherent light from a flashlight, one would measure what seems to be a sub–5-fs pulse. Linear interferometry as described so far provides only the **coherence length** of light, but not its pulse duration. The reason why a linear interferometer is not sensitive to pulse duration is that it compares photons that traveled in one arm with photons that traveled in the second arm (Figure 8.1). If dispersion on both arms is equal, then a minimum coherence length is measured. This observation led to the development of white-light interferometry to measure dispersion in optics. More over, the Fourier transform of the time-resolved linear interferometric scan recovers the spectrum of the pulse.

In order to make interferometers useful for pulse measurement purposes, it is important to measure a nonlinear optical signal that depends on the coherence and pulse duration of the source. The most widely used optic for this task is the second

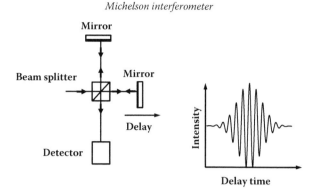

FIGURE 8.1   Interferometer and a linear interference pattern.

harmonic generation (SHG) crystal discussed in Chapter 5. When pulses from each of the arms in an interferometer are recombined and reach the SHG crystal, they convert some of the light into frequency-doubled light. For example, an 800-nm pulse is converted to a 400-nm pulse. The output from the interferometer will then contain frequency-doubled light from each pulse independently and light that depends on the overlap between the two pulses, this is known as an interferometric autocorrelation. When the two pulses are crossed at an angle, the light resulting from the overlap from the two pulses is observed in between the output from the independent pulses. This is what is known as a non-interferometric autocorrelator (AC). Such a device provides a background-free measurement of the pulses in the time delay (Figure 8.2). The time being determined by the distance scanned by the adjustable arm of the interferometer and knowledge of the speed of light is approximately 0.3 μm/fs.

One drawback of the non-interferometric AC is that the measured pulse duration is highly dependent on the crossing angle, the overlap between the two pulses at the SHG crystal, and the precision with which one arm is delayed from the other. All these parameters can affect the measured pulse duration. When the pulses are combined collinearly, one obtains an interferometric autocorrelation (IAC) (Figure 8.3). At long delay times, such autocorrelation has signal from the incoherent sum

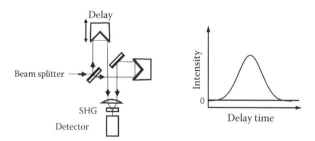

FIGURE 8.2   Non-collinear SHG AC.

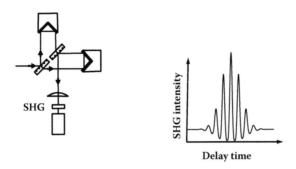

Balanced and symmetrical interferometer

FIGURE 8.3   Interferometric autocorrelation.

of both separate beams. At the time of maximum overlap, it contains signal from the sum of both beams squared, and shortly after, destructive interference from both pulses results in no signal. The interference of both pulses causes modulation at the carrier frequency. The IAC provides several points of reference that are useful for calibrating the measurement itself. First, the signal levels, provided the highest point of the IAC is normalized to one, can ensure the interferometer is balanced and the detectors are acting properly. One expects the signal far from zero delay to reach an asymptotic value of 1/8. Remember that this signal is the sum of the contributions from each of the pulses. The mathematical reason the number is 1/8 is explained as follows: The electric field, of unit intensity, is split into two identical portions (1/2 each). Because intensity is the square of the field, the intensity of each of the two portions is now 1/4. The SHG intensity is proportional to the square of the intensity; therefore, SHG from each portion has an intensity of 1/16. Adding incoherently both portions leads to 1/8 of the intensity: the coherent addition of the two pulses when they are overlapped in time to unity.

The IAC is modulated at the carrier frequency of the pulse, essentially the average optical frequency. This value can be easily obtained using a spectrometer. For example, a femtosecond pulse centered at 800 nm has a carrier frequency that corresponds to an oscillation period of 2.66 fs. The IAC of a Ti:sapphire laser pulse should be modulated at 2.66 fs. The implication is that one can simply count the interference features in the plot to obtain the pulse duration. This is one reason why IAC is a preferred measurement compared to non-IAC. Even if the optical delay is not calibrated, one is able to get a measurement of the pulse duration. When the method provides for an internal reference, it is known as a "self-referencing" measurement.

When the signal measured during an autocorrelation is a nonlinear optical process that can be frequency resolved, for example a **nonlinear polarization**, then one is able to obtain a measurement of the pulse that is both time and frequency resolved. This type of setup is known as frequency resolved optical gating (FROG), and is very useful for pulse characterization. The shortest possible pulses are those when all the frequency components arrive at the same time, and this is easily observed in a FROG trace. Linear chirp is very easy to measure in a FROG trace, especially when the optical non-linearity used to measure the pulses is intensity-induced polarization. Even more popular than polarization FROG is second-harmonic FROG, where the signal detected is the resulting SHG spectrum from the autocorrelation of two pulses (Figure 8.4). In this case, interpretation of the results is no longer straightforward because of the symmetry resulting during the signal generation. For example, both positive and negative chirps look identical in SHG-FROG. The automated analysis of FROG or SHG-FROG traces to retrieve pulse duration and spectral phase information has advanced significantly in the last 15 years. The principle for the analysis is to numerically generate a set of trial spectra and spectral phases of input pulses and to use the information to numerically generate their corresponding FROG traces. Comparison between the numerically generated and experimental FROG trace is used to determine if the numerically generated pulses are a fair representation of the pulse being measured. The algorithm iteratively modifies the numerical pulses until the calculated FROG trace is sufficiently similar to the one experimentally measured. Despite significant progress in the retrieval algorithm, SHG-FROG traces may have ambiguities, this implies that there may be more than one spectrum and phase that represents the actual pulse.

Fourier-transform spectral interferometry (FTSI) takes advantage of spectral interferometry observed in the frequency domain (Figure 8.5). Consider two femtosecond pulses that are replicas except for a time delay shift. When such pulses are collinearly combined and a spectrum is measured, one observes an interference that is caused by the time delay between the two pulses. The time delay introduces an additive phase (i-$\omega\tau$); the fringe period is therefore expected at (2-π/τ). The larger the time delay, tau, the shorter the spacing between the interference fringes. By Fourier-domain filtering, one is then able to retrieve the relative phase between the two pulses. Note that this method is ideal when one of the pulses is known, because it retrieves the relative phase between the two pulses.

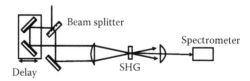

FIGURE 8.4   FROG and SHG FROG.

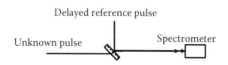

FIGURE 8.5   FTSI.

More importantly, this method does not require femtosecond pulses because it relies on linear interferometry and can be used to characterize the dispersion of an optic or optical setup when compared to a reference path.

The power of spectral interferometry is at the heart of a pulse characterization method known as spectral phase interferometry for direct-electric field reconstruction (SPIDER) (Figure 8.6). To understand this method, we first recall how two pulses displaced in time show interferometric fringes when detected by a spectrometer of sufficient frequency resolution. Similarly, if the two pulses are displaced in space, an interference pattern is observed: this is how shear interferometers work. Shear, therefore, refers to a temporal or spectral shift that causes interference between two pulses. For the purpose of pulse characterization, the interferometry involves two pulse replicas that are spectrally sheared. While there are multiple implementations of the method, the most common involves the creation of the two sheared replicas through up-conversion with a highly chirped pulse. The reason for the highly chirped pulse is that during the up-conversion process, each of the pulse replicas is up-converted by essentially monochromatic fields with a fixed frequency difference. The frequency difference causes the shear that then leads to the interference between the up-converted replicas. As in all interference-based methods, the magnitude of the shear determines the resolution.

So far, the methods described for pulse characterization are based on interferometry. One of the caveats of both time and frequency interferometric pulse characterization methods is that they work well for coherent laser pulses but not for incoherent pulses. This observation is particularly important when trying to characterize the output of an ultrafast laser source whose output varies shot-to-shot in either spectrum, spectral phase, or pulse shape and one is trying to measure the average pulse duration based on a measurement that samples two or more pulses from the laser output. The problem arises from the fact that the interferometric result contains a representation of the pulse that is primarily based on the spectral bandwidth of the pulse. If, however, the output of the laser being characterized

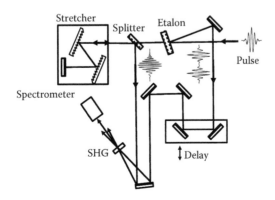

FIGURE 8.6   SPIDER.

has major shot-to-shot variations as indicated above, the interferometric charac-terization will result in a much shorter estimate of the average pulse duration of the laser pulses. Information about the true nature of the laser output is avail-able in the interferometric-based methods; it appears in the wings or pedestal of the autocorrelation function or on the reduced fringe depth of the spectral-shear-based measurement. Unfortunately, it takes a high degree of skill to obtain inter-ferometric measurements without any pedestal or with 100% modulation depth in the fringes. As we shall see in Chapter 10, other methods for pulse character-ization are more amenable to measure the output of ultrafast laser sources with pulse-to-pulse fluctuations. Before we review those methods, we need to look more deeply into how phase and bandwidth determine pulse characteristics.

9 Shaping the Laser Pulses

In Chapter 3, we described transform-limited pulses as those where all the frequencies within the spectrum of the pulse arrive at the same time. In Chapter 4, we described how, when transform-limited pulses go through media, they acquire a phase that leads to pulse broadening. In this chapter, we provide a more thorough discussion of how phase affects the temporal properties of a laser pulse. For simplicity, we shall assume that the spectrum of the pulse has a Gaussian shape; however, the concepts discussed here apply in general to any spectral shape. Our discussion about phase will focus on the spectral phase function, which is an expression that describes how the phase varies as a function of frequency. First, we can start with the case where the spectral phase function equals zero. In this case, all frequencies have the same phase and therefore interfere constructively to produce the shortest possible pulse based on the given spectrum. Note, at this point, that adding a constant phase across all frequencies does not affect the pulse characteristics. Therefore, we can quickly conclude that the absolute phase of the pulse is not important in determining the duration of a laser pulse. In the case of pulses having a pulse duration that is shorter than three optical cycles, the absolute phase determines where in the amplitude envelope one observes a maxima and minima in the electric field. Controlling the absolute phase has allowed scientists to control the direction at which electrons are ejected from atoms. The majority of femtosecond lasers, however, have pulse durations that range from 10 to 100 optical cycles in duration, and in these cases, changes in the absolute phase make little or no difference.

Beyond the absolute phase, the most interesting case to consider is the effect of a spectral phase function corresponding to a linear change with respect to frequency. As a pulse of light evolves in time, each frequency component advances according to the product between the frequency and the time interval. Therefore, if the slope of the linear spectral phase is equal to 100 fs, the resulting pulse corresponds to one identical to the one when the spectral phase function was zero but delayed in time by 100 fs (Figure 9.1). When the linear spectral phase function equals −50 fs, the pulse now corresponds to one that has been advanced by 50 fs. We note here that the pulse characteristics, in terms of pulse duration, intensity, and spectrum, remain identical; therefore, pulse characterization methods do not detect the absolute position in time of the pulse—neither does the overall linear spectral phase of the pulse. The reason why the linear spectral phase is the most interesting is that every other spectral phase function can be decomposed into linear portions and understood without the need to resort to the mathematical functions, which, upon Fourier transformation, yield the time-domain representation corresponding to the pulse with an arbitrary spectral phase.

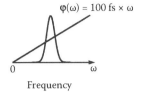
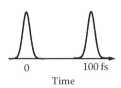

Frequency

Time

FIGURE 9.1 How a linear phase corresponds to a time delay.

Based on information in the previous paragraph, we are now ready to explore how different spectral phase functions affect laser pulses without the need of Fourier transforms. We will continue this discussion considering the case in which the pulse acquires a quadratic spectral phase function. While the choice of function may seem arbitrary, we note that the refractive index of most transparent materials follows a frequency dependence that can be decomposed into an absolute value, followed by a linear slope, and then a quadratic dependence on frequency. Given that the absolute value and linear dependence have no effect on the pulse, we are left with the quadratic dependence as the major influence on the pulse. We now consider a positive quadratic spectral phase to be a combination of positive slopes from the vertex to higher frequencies and negative slopes from the vertex toward lower frequencies. One can now deduce that a positive quadratic spectral phase results in a pulse whose higher frequencies are delayed and lower frequencies are advanced. Interestingly, the degree of stretching of the pulses is linear with frequency, i.e., increases linearly from the vertex; therefore, the pulse is evenly stretched in pulse duration. We now note that a quadratic spectral phase corresponds to linear chirp as discussed in Chapter 4, also known as linear dispersion (Figure 9.2). Clearly, the magnitude of the quadratic spectral phase determines the extent to which the pulse is stretched in time, and the sign determines if frequencies are advanced or delayed. It may seem that the spectral position of the vertex is important, but it is not. The reason why the position of the vertex is not important is that the vertex position controls only the time of arrival of the pulse. This is because addition of a linear function to the quadratic phase

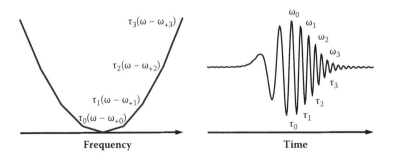

FIGURE 9.2 Chirp in terms of linear functions.

function displaces the vertex position. We have already remarked that a linear spectral phase function does not affect the pulse in other ways other than its time of arrival.

The next spectral phase function that will be discussed is known as third-order dispersion and corresponds to a cubic spectral phase (Figure 9.3). In this case, a positive cubic spectral phase function causes the delay of frequencies that are higher and lower than the frequency at the inflection point. The pulse resembles one that is wind-swept toward longer times. A pulse with positive third-order dispersion has a fast rise and a slow decay, displaying interference between higher and lower frequencies in the tail. The magnitude of the third-order dispersion determines the extent of stretching of the pulse and the sign determines if the pulse experiences stretching toward longer or shorter times.

Having discussed quadratic and cubic spectral phases allows us to consider higher-order dispersion. For all even functions, one observes pulse broadening similar to that caused by chirp; however, the pulse stretching increases nonlinearly with frequency. For all odd functions, one observes one-sided pulse broadening similar to that caused by third-order dispersion; however, the pulse stretching increases nonlinearly away from the main part of the pulse.

Following the discussion above, the reader should have sufficient information to infer how an arbitrary phase may affect a given pulse. The key is to decompose the arbitrary phase into linear components. Determining accurate pulse durations and temporal pulse shapes following arbitrary phase modulation requires mathematical rigor. Fortunately, FemtoPulse Master originally introduced by Biophotonic Solutions Inc., a free program available for download, does this (Figure 9.4). If the reader is interested in obtaining a copy of this software, search for it online or request a copy from the author.

Working with FemtoPulse Master is simple. The program starts with certain assumed pulse characteristics such as transform-limited pulses that are centered at 800 nm with 50-fs pulse duration. The user can then manipulate the phase using functions or arbitrarily by sliding the phase of frequencies within the spectral

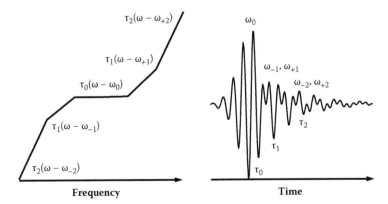

FIGURE 9.3 Cubic phase in terms of linear functions.

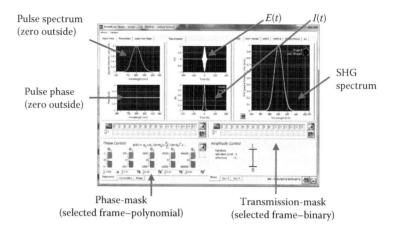

FIGURE 9.4   FemtoPulse Master front panel.

bandwidth of the pulse. The user can choose to look at how the pulse changes in time through selection of a pulse characterization method, or by tracking how the introduced phase affects how the second harmonic spectrum changes. It is up to the reader to play with FemtoPulse Master and confirm everything discussed so far about how the spectral phase affects the output pulses of a femtosecond laser.

The discussion in this chapter is valuable (a) because most optics affect spectral phase and (b) because obtaining short pulses requires us to mitigate unwanted dispersion. As indicated earlier, transmission through any transparent medium other than vacuum results in dispersion. The main component of dispersion is a quadratic spectral phase; however, third-order and fourth-order dispersion and even higher orders of dispersion affect the pulses. The higher-order terms arise from either propagation of the pulse through a large path in dispersive media, from spectral phase variations in the reflective and anti-reflective coatings on optics, and finally from nonlinear optical effects along the path of the pulses. In principle, pulse characterization should yield the spectral phase of the affected pulse, and pulse shaping should allow one to recover transform-limited pulses. Here, pulse shaping may simply mitigate dispersive effects using static optics, or alternatively, it may involve a programmable pulse shaper that allows the user to introduce a spectral phase that cancels any spectral phase distortions in the pulse.

In general, pulse shaping involves the control of the spectrum or spectral phase of a laser pulse. This is why another expression for pulse shaping is spectral filtering. Ultrafast pulses, because of their broad bandwidth, are affected by transparent media including air through dispersion. In addition, most types of dielectric coatings and some types of metallic coatings lack the ability to support the entire bandwidth of the pulses and therefore act as passive pulse shapers. Dispersion is one of the best-known effects on femtosecond pulses. Interestingly, over a decade before femtosecond pulses were produced, Edmond Treacy set out to develop optical setups involving prisms or gratings to mitigate dispersive effects. Treacy realized that femtosecond pulses would need to be compressed after they had suffered dispersive broadening. Treacy's designs are still used in most femtosecond lasers. The rationale for his designs was to create optical layouts where higher frequencies travel a shorter path than lower frequencies. When done properly, the linear advancement of higher frequencies cancels the linear delay of such frequencies caused by dispersion. As discussed earlier, dispersion may have higher-order components such as cubic, quartic, and so on. It is possible to combine gratings with different properties or gratings and prisms in order to control dispersion including third-order dispersion. More recently, dielectric coatings involving tens of layers of alternating optical materials have been designed to introduce controlled amounts of second- and third-order dispersion. Femtosecond pulses can be compressed using such so-called chirped mirrors, provided the laser pulses are reflected a sufficient number of times by the optic.

As long as nothing changes in the laser system, optical setup, or experiment, the static pulse compression is very practical. However, when the laser system changes in alignment or internal power, the optical setup changes in distances or optics, or when the experiment changes, compression parameters change. Optimization of static optics requires accurate pulse characterization, and the process of pulse compression can become time consuming and tedious. Therefore, the goal is to reduce dispersive effects to a degree where their effect is assumed negligible on the measurements being performed. This implies that most experiments are not carried out with transform-limited pulses.

Programmable pulse shapers are instruments capable of controlling the spectrum and spectral phase of femtosecond pulses and can therefore be used to mitigate all orders of dispersion (Figure 9.5). Pulse shapers have optics that separate the different frequency components, such as gratings, and have a programmable element, such as a spatial light modulator (SLM), that can modify the phase and amplitude of the separate frequency components of the pulse before the pulse is reconstructed at the output of the pulse shaper. Amplitude modulation can be used to attenuate parts of the laser pulse spectrum in order to produce a desired spectral shape. For example, producing a Gaussian spectrum ensures that the compressed pulses will have a Gaussian shape in the time domain as well. Note that pulse shapers cannot add spectral intensity; therefore, the bandwidth of the input pulse determines the frequency range available for pulse shaping. In terms of phase control, pulse shapers can introduce an arbitrary phase to portions of the

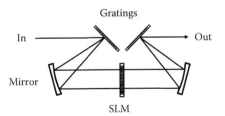

FIGURE 9.5 The 4f pulse shaper.

spectrum. The spectral resolution of the pulse shaper is a parameter that depends on the pulse shaper design and the input spectrum. In general, the resolution is either equal to the pulse bottom-to-bottom bandwidth divided by the number of individually controlled elements of the programmable device (typically 128, 640, or 800). While devices with a larger number of pixels exist, and other shaper designs exist where the phase control element does not have pixels, the spectral resolution stated above still holds. There are designs that permit even higher spectral resolution; however, such designs introduce greater loss in the optical setup. In conclusion, pulse shapers provide the means to control the spectrum and spectral phase of femtosecond pulses; unfortunately, one needs to know precisely the spectrum and spectral phase of the input pulse and the dispersion of subsequent optics in order to ensure that the femtosecond pulse reaches its target with a perfectly determined spectrum and spectral phase. In the following two chapters, we discuss how to accomplish this task.

10 How to Measure Femtosecond Pulses (Part 2)

From the previous chapter, we learned that delivering femtosecond pulses from the laser to the desired target requires some form of pulse characterization and some form of pulse shaping. This is because the pulses from the laser are never perfect and because the pulses acquire additional dispersion as they reflect from mirrors and transmit through optics. In general, industrial applications and simple experimental setups rely on mitigation of dispersion to second order using a prism pair or grating pair and track a nonlinear optical signal such as second harmonic generation in order to achieve pulses that are between 10% and 20% from being Fourier transform limited. However, demanding applications require reliable and reproducible delivery of pulses with precise spectral phase and amplitude characteristics every day. In this chapter, I describe how a pulse shaper can provide both functions, accurate pulse characterization and pulse compression. Moreover, the pulse shaper can be fully automated, simplifying the consistent delivery of transform-limited (or precisely shaped) pulses at the target.

In Chapter 8, when we discussed pulse characterization, it became clear that interferometry featured prominently in all forms of pulse characterization. Interferometry takes advantage of constructive and destructive interference of light waves to measure the spectral phase of the pulses. Furthermore, we discussed how linear interference is not very useful for pulse characterization because it reflects properties of the pulses that are related purely to the spectrum and not the spectral phase of the pulse. Therefore, it is important to introduce the concept of **nonlinear optical interference**.

Consider the nonlinear optical process of second-harmonic generation or, in general, sum-frequency generation. This multiphotonic process involves the sum of the energies of two photons (input) to produce a third photon (output) with energy equal to the sum of the two photons. The phase of the output photon equals the sum of the phases of the input wavelengths. Assume the input pulse is centered at 800 nm. Note that 400-nm photons can be generated from the sum of two photons with 800 nm wavelength (recall that photon energy is inversely related to the photon wavelength). Similarly, 400-nm photons can also be generated from the sum of one photon with 795 nm wavelength and one photon with 805 nm wavelength. The phase of the 400-nm resulting photon equals the sum of the phases of the input wavelengths. When the input pulse is transform limited (all photons having the same phase), then the 400-nm photons from the two different combinations (2 × 800 and 795 + 805) of photons have the same phase and add up constructively $(1 + 1)^2 = 4$. Constructive interference would cause the output to be the sum of the two fields squared. This can be drastically changed if the phase of the 795-nm photon, for example, is changed to be out of phase with that of the 800-nm and 805-nm photons. Now, the resulting 400-nm photon from the sum of (795 + 805) is out of phase with the 400-nm photon produced by the 800-nm photons. The two sources of 400-nm photons interfere destructively $(1 - 1)^2 = 0$ and no photons emerge at 400 nm. Controlling nonlinear optical interference through manipulation of the spectral phase of the pulse is a process known as multiphoton intrapulse interference (MII) (Figure 10.1).

The realization that second-harmonic generation, sum-frequency generation, difference-frequency generation, and essentially most nonlinear optical processes can be controlled by the spectral phase input because of MII has very important implications. The reader will recall the discussion on pulse characterization. All conventional methods require interferometric setups of two or more pulses. The setups involve multiple paths, with adjustable lengths and overlapping in space and time of two or more beams. Typically, two pulses that may be invisible to the naked eye need to be overlapped in three dimensions within a micron. The complexities result in uncertainties and noise in the data acquisition process, and this affects the reliability of the retrieved laser pulse parameters. If, on the other hand, one could take advantage of the MII process to characterize a laser pulse, that would eliminate the need for splitting and recombining pulses after independent optical delay lines. Furthermore, MII-based pulse characterization would provide a direct determination of the spectral phase of the pulse instead of an indirect process based on retrieval algorithms.

Nonlinear optical processes are sensitive to spectral phase; this has been known for decades. However, it was also realized that the dependence was not sufficient to provide robust pulse characterization. For example, the dependence of second harmonic generation (SHG) on chirp is used to optimize approximately

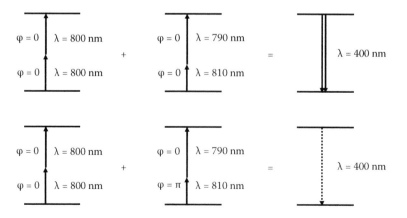

FIGURE 10.1 Concept of multiphoton intrapulse interference (MII).

a simple prism- or grating-based compressor. Robust pulse characterization, however, requires the elimination of ambiguities. As we shall see, this is typically accomplished by making two or more measurements using different reference phases as explained next.

Consider an input femtosecond pulse with an unknown spectral phase that is causing pulse broadening and the appearance of a pedestal. At our disposal, we have a grating-based compressor, an SHG crystal, and a spectrometer. Our goal is to find the spectral phase function that is causing temporal distortions on the pulse. Given that the absolute phase and linear phase functions do not affect the temporal properties of the pulse, we can dismiss those. At this point, we introduce some mathematics to describe a function that approximates the phase distortions as a polynomial equation. We then introduce some calculus to reduce the polynomial equation by two orders. The reader may want to skip this level of detail, or enjoy the detail and wonder why this approach was not developed 20 years earlier.

We can think of the unknown phase distortions as being described by a polynomial equation with terms that have powers of the wavelength going from zero (affecting all wavelengths equally), to linear, quadratic, cubic, and so on. Given that the first two terms have no effect on pulse duration, they are not relevant to pulse characterization. Pulse characterization in a sense is a process to measure the terms starting from the quadratic power and higher orders. In calculus, there is a function that reduces the power of a polynomial equation; it is known as the derivative. The derivative of the quadratic term is a linear term, and the second derivative of the quadratic term is a constant. Therefore, when one takes the second derivative of the unknown phase, one eliminates lower-order terms and the quadratic term is observed as a constant. Once the second derivative of the unknown phase function is determined, the spectral phase can be obtained through numerical integration, a process that is very robust with respect to noise.

Based on the previous paragraph, we can define the challenge of pulse characterization as that of finding the second derivative of the spectral phase as a function of wavelength. Experimentally, this turns out to be quite simple. We take advantage that we know that when the second derivative of the spectral phase (local chirp) is zero, nonlinear processes such as SHG are maximized. This is where having adjustable pulse compression optics such as a pair of gratings or prisms becomes practical. By scanning the compressor from negative to positive chirp, we change the curvature of the phase. At some values of chirp, the unknown spectral phase will be cancelled by the reference chirp introduced. When we record SHG spectra at each value, we find maxima in the spectrum that correspond to positions where the local chirp becomes zero. As we gather all the SHG spectra as a function of input chirp, we obtain an MII-phase scan or MIIPS trace that can be used to determine the spectral phase of the pulses (Figure 10.2).

MIIPS pulse characterization is illustrated by considering first three types of pulses: transform limited, linearly chirped, and pulses with third-order dispersion. For transform-limited pulses, as we scan chirp, we find that maximum SHG takes place for all wavelengths at the same value of reference chirp, where the chirp introduced equals zero (see Figure 10.3). Now, let's consider the pulse has

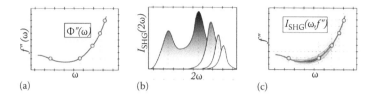

(a)   (b)   (c)

FIGURE 10.2 Concept of MIIPS-2. (a) Consider a pulse that is broadened because of a spectral phase that has non-zero curvature. The second derivative of such an arbitrary spectral phase is depicted as the red curve being plotted as a function of frequency. The goal of pulse characterization is to determine the spectral phase experimentally (to measure the position of the green dots). (b) By introducing to the pulse predetermined amounts of chirp, which correspond to constant displacements in the second derivative space (panel a), and acquiring SHG spectra of the pulses for each case, one obtains the position of the green dots that correspond to local maxima. (c) When all the SHG spectra are collected as a contour map, one obtains directly the second derivative of the unknown spectral phase.

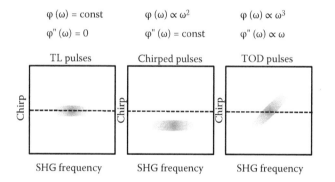

FIGURE 10.3   MIIPS-2.

acquired some amount of positive linear chirp. In this case, we find that maximum SHG takes place for all wavelengths at the same point, this time a point that corresponds to an amount of negative chirp that compensates for the positive chirp of the input pulses (Figure 10.3, middle). As a third example, we consider a pulse that has an amount of positive third-order dispersion. In this case, the maximum SHG signal is obtained at different wavelengths depending on the reference chirp. Interestingly, as we look at all the SHG spectral intensity as a function of reference chirp, the signal forms a diagonal feature. At this point, it is worth remembering that MIIPS measures the second derivative of the phase function. The second derivative of a cubic variable is a linear variable. This is why the third-order dispersion becomes a linear function with a slope proportional to the amount of third order dispersion in the pulse in a MIIPS measurement. Beyond the simple functions considered here, MIIPS measures the second derivative of any complex function, as shown in Figure 10.2.

Once the second derivative of the phase for a pulse has been measured, one is able to calculate by numerical integration the phase of the pulse as a function of wavelength. This information, together with the spectrum of the pulse, is sufficient to characterize the pulse spectrally and temporally.

In the previous example, a compressor in the femtosecond laser was used to introduce linear chirp. The linear chirp can be considered to be the "reference" phase that is used for measuring an unknown phase. Once the pulse has been characterized, one may want to reduce or completely eliminate phase distortions in order to achieve the shortest possible pulses at the intended target. If the phase distortion is only linear chirp, then the compressor can be adjusted to obtain transform-limited pulses. However, if the pulses have distortions involving third- and higher-order terms, then it is extremely difficult to compress those using static optics. Arbitrary phase correction requires an adaptive pulse shaper such as those described in Chapter 9.

Turns out that the distance between the gratings in the compressor introduces linear chirp without other distortions, provided it is well aligned. However, for high-accuracy characterization, it is possible to achieve higher-accuracy pulse characterization using MIIPS.

Consider a pulse shaper, such as those described in Chapter 9, is available. In this case, the shaper can be used to introduce the reference phase with much greater accuracy. The pulse shaper can introduce a linear chirp to characterize the pulses as described above. The pulse shaper can now correct the phase distortions measured. After correction, one can run the MIIPS measurement once again and confirm all phase distortions have been eliminated and transform-limited pulses have been achieved. This process works very well, in fact one is able to achieve correction that is within a percent of the theoretically shortest pulses possible given the input bandwidth.

Scientists, in particular, are not happy leaving a few percent phase distortions uncorrected. We have found that it is advantageous to use a sinusoidal reference phase in order to achieve even greater accuracy when using MIIPS, especially after linear chirp has been used already to remove large phase distortions. Here we consider the case in which a sinusoidal phase is scanned. In this case, one obtains diagonal SHG features that isolate spectral regions of the pulse. One can think of this process as follows. The sinusoidal function introduces either positive or negative chirp except at the inflection point where it introduces a linear phase that does not affect SHG intensity. As the sinusoidal phase function is scanned across the spectrum using the pulse shaper, so that the inflection point samples all the wavelengths, one obtains diagonal features that repeat every π, provided the pulses are transform limited. Deviations from parallel lines separated by π indicate phase distortions, which are measured and corrected as described earlier, except that one needs to remove the sinusoidal dependence of the reference phase. Combining chirp and sinusoidal MIIPS, one is able to obtain pulses that are within 0.01% or better of the theoretical value (Figure 10.4). These levels of accuracy cannot be achieved without a pulse shaper and a MIIPS-like method for pulse characterization.

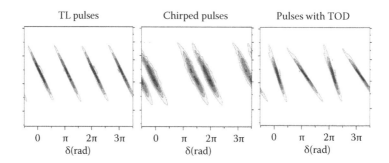

FIGURE 10.4   Sinusoidal MIIPS for TL pulses, for linear chirp, and for pulses with third order dispersion.

A final word about MIIPS—the method does not require overlap between pulses, the pulse shaper operates as a common path interferometer, the method is self-referenced, and it is possible to know that transform-limited pulses have been achieved even if the pulse shaper is not well calibrated. Altogether, it is a powerful platform for pulse characterization and adaptive compression that is capable of controlling the delivery of transform-limited pulses or precisely shaped pulses to the target. In the following chapters, we explore applications of a femtosecond laser source with MIIPS adaptive shaping. Finally, in Chapter 17, I describe how shaper-based automated pulse compression allows the scalable production of femtosecond lasers, and makes them fully computer controllable. These properties are key for future price reductions in the cost of femtosecond lasers by one and, later, two orders of magnitude.

11 Applications of Shaped Pulses to Biomedical Imaging

Finally, we get to the chapters that discuss application of femtosecond pulses. For the readers that have enjoyed or endured the previous chapters, I hope you learned from the background; you are now well equipped to understand the versatility of femtosecond lasers. For the readers that skimmed or altogether skipped the technical chapters, if you encounter some undefined technical term, feel free to find an explanation by following the index at the end of the book or by taking advantage of online information.

This chapter discusses the application of shaped laser pulses for microscopy and, in particular, biomedical imaging. To some extent, we are considering an extension of the light microscope. However, the goal is not simply to replace it. The goal is to construct a novel type of microscope capable of obtaining depth resolved images in vivo and to provide chemical information with subcellular resolution. The powerful new microscope would then be automated so that it can be used in the clinic to identify and diagnose multiple different types of cancers. Such a microscope would help determine if surgery is required based on an optical biopsy. If surgery were indicated, the microscope would then be used to determine the locations of the tumor margins. The same microscope would be able to provide depth-resolved images of a living brain. If the microscope is designed at the tip of an endoscope, it can provide cancer diagnosis of internal organs such as the colon and esophagus. Many scientists have been working towards the microscope being described here; however, realizing a practical system that achieves high-resolution depth-resolved images with rich chemical content has been a challenge. Here, I describe how shaped femtosecond pulses are addressing these needs.

When a laser replaces light, one is able to achieve higher intensities and greater rejection of scattered light. The introduction of femtosecond laser pulses, however, makes accessible a number of nonlinear optical processes that can be used for contrast. More importantly, because the signals being detected are nonlinear, they emanate primarily from within the focal region where the peak intensity is the highest. Typically, this region is about 1 µm deep and is diffraction-limited on the plane of excitation. Obtaining depth-resolved images becomes possible by systematically moving the microscope objective toward the sample in order to move the focal plane deeper into the object being imaged. Once femtosecond laser pulses became commercially available in the 1990s, a number of nonlinear optical microscopies were demonstrated. Among the most practical and widely known is two-photon excitation fluorescence (TPEF) microscopy. Three- and even four-photon excited fluorescence microscopy have been demonstrated since then. TPEF takes advantage that a wide number of fluorescent compounds absorb near 400 nm light. When two photons of 800 nm light from a titanium–sapphire femtosecond laser impinge on the molecule, the molecule absorbs the equivalent of one 400-nm photon and fluoresces in the visible spectrum. The fluorescence collection filter rejects laser light, leaving only the visible light to go through. The use of multiple different fluorescent compounds that are designed to label different parts of the cell can lead to impressive multicolor images. In some cases, the fluorescent compounds are introduced through genetically encoded fluorescent proteins, and in some cases, the fluorescent compounds are naturally found in the tissue.

Early efforts on pulse shaping and biomedical imaging focused on reducing the damage induced by the laser pulses on the tissue being imaged. In one approach, pulses were split into multiple replicas of lower intensity. Those types of approaches don't bring about new functionalities. Here, we consider how shorter pulses can accomplish both less photodamage and new capabilities as described below.

There are other nonlinear optical sources of contrast such as second harmonic generation (SHG, discussed in Chapter 5) and third harmonic generation (THG). Most highly structured biological tissue, in particular protein fibers, has a certain symmetry property described as having no that lacks what center of inversion. Certain types of tissues, which include skin, muscles, and bones, are materials that contain such inversion lacking materials such as collagen, ellastin, tubulin, and generate significant SHG signals. Biological materials with refractive indexes very different from that of water, such as lipids, are good sources of THG, which requires an index of refraction discontinuity at the focal plane to become visible. Both SHG and THG have been used for biomedical imaging. The ultimate microscope is one capable of inducing and detecting all these nonlinear optical signals.

There is one more modality worth discussing here. The modality is related to molecular vibrations. Given that different molecules have different vibrational signatures, vibrations can also be used to identify among different tissues. Approaches that take advantage of molecular vibrations include coherent Stokes and anti-Stokes Raman scattering (CARS and CSRS). In these modalities, the laser induces vibrations that can be used to identify different compounds such as lipids, water, or proteins. Broad-bandwidth femtosecond pulses are particularly efficient at inducing coherent vibrational signals, thereby providing robust CARS or CSRS signals for microscopic applications. The microscope equipped for this modality can collect a vibrational spectrum every pixel, or it can use the pulse shaper to restrict the signal to a single vibrational signature. The latter approach is more amenable for fast imaging because one is not limited by the time required to acquire a full spectrum.

Let's make a summary of what we are describing: A microscope that uses femtosecond pulses that can excite multiple different fluorescent compounds through two- and three-photon excitation (2PEF and 3PEF), that can identify tissues by their symmetry (SHG) by their index of refraction (THG), or by their vibrational signatures (CARS and CSRS), see Figure 11.1. The instrument being envisioned and recently demonstrated is incredibly powerful. It can be used by a number of physicians to aid in their diagnostic work provided the microscope is affordable and can

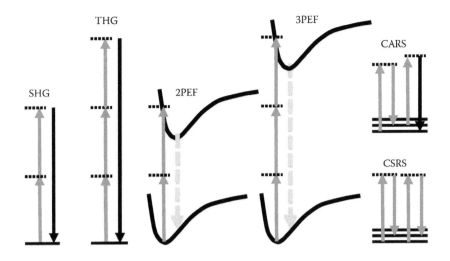

FIGURE 11.1 Nonlinear optical signals available for multimodal imaging with broad bandwidth femtosecond laser pulses.

be operated simply, not requiring a PhD in optical sciences to make it work. Finally, it is very important to make sure that the femtosecond laser pulses themselves are not damaging the tissue being imaged.

The first requirement for such a microscope is short laser pulses. How short? This is a question that has been asked for many years. It is useful to recognize that TPEF and SHG signals follow an inverse linear relationship with pulse duration. Therefore, pulses that are 10 times shorter lead to 10 times brighter signals (Figure 11.2). THG and three-photon excitation fluorescence follow an inverse-squared relationship with pulse duration. Therefore, pulses that are 10 times shorter lead to 100 times brighter signals. Clearly, shorter pulses seem to have significant advantages. In fact, while early application of multiphoton microscopy made heavy use of fluorescent staining, modern shorter pulsed lasers make unstained biomedical imaging possible because of the increased multiphoton excitation efficiency.

Historically, femtosecond laser systems capable of generating pulses shorter than 100 fs were not readily available. Furthermore, high-order dispersion (Chapters 3 and 4) introduced by the microscope objective prevented delivery of pulses much shorter than 50 fs. Some of the early experiments with significantly shorter pulses seemed to indicate that it was easier to damage biological tissue with shorter pulses. It is not difficult to understand how pulses that are 10 times shorter have 10 times higher peak intensities. It was then realized that shorter pulses, which already create much brighter images, should be attenuated to prevent tissue damage.

The difficulties posed by the early experiments with sub–100-fs pulses led to the development of lasers with pulse durations as long as 200 nm tunable over a broad range of wavelengths. Such lasers have been extremely popular because

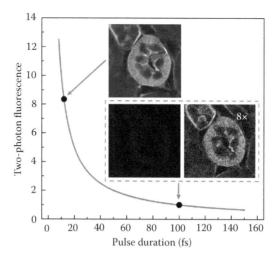

FIGURE 11.2   Two-photon imaging with 12-fs pulses results in 8× brighter images when compared with 100-fs pulses.

their narrow spectral bandwidth <10 nm makes them less susceptible to dispersion. Unfortunately, the pulse duration is so long that one needs to use powers that exceed what has been determined to be safe for in vivo imaging on humans.

The introduction of broadband femtosecond lasers capable of supporting 10-fs pulse durations and even shorter, together with adaptive pulse shapers for pulse compression that can handle high-order dispersion automatically, has renewed the question about the merits for shorter pulses (Figure 11.3).

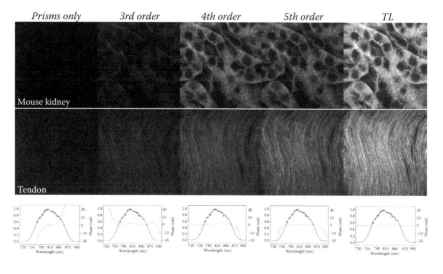

FIGURE 11.3   Two-photon fluorescence (top) and SHG (bottom) imaging as third- and higher-order dispersion of the laser pulses are corrected.

Given that SHG and THG are parametric processes that are wavelength independent, the discussion provided here centers on TPEF. Ideally, one would like to be able to excite a wide range of fluorescent compounds. For those that use titanium–sapphire lasers, excitation at 750 and 850 nm seems to be very important. This implies that a 100-nm bandwidth centered at 800 nm is already quite good. A bandwidth that goes from 700 to 900 nm would be ideal; such a bandwidth (full width at half maximum) would support pulses that are approximately 5 fs in duration. Such short pulses require adaptive pulse compression and could excite compounds with equivalent one-photon absorption ranging from 350 to 450 nm. The wavelength range includes most endogenous biological fluorophores as well as all the fluorescent proteins used for imaging. These lasers and pulse shapers are already commercially available; however, they are expensive.

Would shorter pulses be better? One can answer this question from different points of view. From a practical point of view, pulses shorter than 10 fs pose major difficulties because optics such as microscope objectives and associated microscope filters were not designed to handle such broad bandwidths. Optical microscopes depend on the optical design of the microscope objective, a complex optic consisting of multiple coated lenses. Such optics are able to correct for spherical and chromatic aberrations; however, they do so at the cost of substantial spectral dispersion. While adaptive pulse compression has allowed for delivery of sub–10-fs pulses and perhaps pulses as short as 6 fs, shorter pulses have not been used for biomedical imaging.

There are different arguments, however, that make shorter pulses desirable. First, broader bandwidth pulses are able to address a larger class of chromophores. In this case, an octave spanning pulse, which is equivalent to 3.5-fs pulses centered at 1060 nm, would be able to excite any chromophore by two- or three-photon excitation. Note that one does not need to excite them all at once. It is possible to introduce phase functions to selectively excite in a certain central two- or three-photon frequency. In Chapter 9, we discussed shaping of laser pulses. Through phase modulation and taking advantage of multiphoton intrapulse interference, one can use a broadband pulse that is shaped to selectively two-photon excite molecules in a very narrow spectral bandwidth.

An additional reason why broad-bandwidth lasers are desirable is that one can use them for driving molecular vibrations. When the bandwidth of the laser pulse is broad enough to encompass the energy of a vibration, the pulse can drive the vibration coherently. Coherent vibrational excitation, when all the molecules of a certain type are vibrating at the same time and at the same frequency, leads to very intense scattering signals. Therefore, it is possible to build a microscope for exciting and detecting vibrational excitation. The advantage of such vibrational microscopy is that each type of molecule has a set of unique vibrations. A vibration-selective microscope can thus be used for chemically selective imaging, without the need for introducing special fluorescent labels into the sample or tissue. Vibrational microscopy, which started with infrared light and then evolved as spontaneous Raman microscopy, can now be accomplished with greater sensitivity and speed by taking advantage of coherent vibrational excitation. The most advanced method being used at present is known as femtosecond coherent anti-Stokes Raman scattering (CARS).

CARS, as it is being used for microscopy, involves the impulsive excitation of molecular vibrations followed by a narrow band probing laser. The molecular vibrations modulate the probing laser and that modulation is observed as a vibrational spectrum when collected in a spectrometer. When implemented as such, one obtains a vibrational (Raman) spectrum at each pixel. Advances in high-speed array detectors have made it possible to greatly speed the acquisition of CARS images. In terms of molecular vibrations, they range in energy from 300 to 3000 cm^{-1}. Being able to excite coherently the higher-energy vibrations, which include the C–H stretching motion that is very prevalent in fatty molecules, requires pulses as short as 7 fs.

In summary, femtosecond laser pulses are revolutionizing biomedical imaging. One is already able to see some early publications in which broadband pulses are used to collect four different modalities of excitation and at the same time two different vibrational signals. When such a microscope was used to image unstained breast cancer histology slides, it was able to identify the location of water and lipids through their vibrational signatures. Collagen generated strong SHG signals. In addition, new markers for metastasis were found through three-photon fluorescence. One can envision that, in the future, unstained multimodal biomedical imaging will become much more common. This progression will hinge on two technical developments. First, the continued evolution of femtosecond lasers. For biomedical imaging, one will require robust and low-cost fiber lasers producing pulses with wavelengths between 1.0 and 1.8 μm.

The reader may note that the wavelengths being envisioned here are longer than the 800 nm discussed so far. Longer wavelengths penetrate deeper without significant scattering, longer wavelengths cause less photodamage, and, finally, longer wavelengths can be used to achieve two-, three-, and four-photon excitation. The most striking demonstrations of how longer-wavelength pulses are better than their shorter-wavelength counterparts come from depth-resolved brain images where subcellular resolution three-photon fluorescence images were obtained 2 mm from inside a living brain. The second technical development needed for the routine application of such lasers will be adaptive pulse compression so that the user of the microscope can dedicate to imaging and not to characterize and compress laser pulses. Efforts to develop the needed technology required to transition from academic laboratories to industrial research and development, are already underway. In parallel, the application of multimodal imaging for the identification of cancerous tissues in excised tumors is already being evaluated and may become approved for medical diagnosis as the technology becomes commercially available (Figure 11.4). The images shown in Figure 11.4 were obtained in the laboratory of Prof. Stephen A. Boppart.

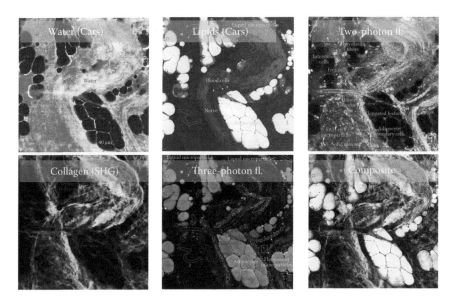

FIGURE 11.4 Multimodal imaging of a histological slide from a breast cancer tumor.

Once the microscope being envisioned here becomes available, it will have one additional capability that will make it highly desirable. Once cancerous tissue is identified, the laser intensity can be increased and the same laser and setup can be used to kill the cancerous cells. The potential diagnosis and therapy or theranostic properties of femtosecond lasers are just beginning to be explored. I believe they have a bright future.

12 Applications of Shaped Pulses to Standoff Detection of Explosives and Other Materials

This chapter addresses the challenging task of detecting explosives and other materials from a safe distance. In essence, the challenge can be described in general terms as being able to detect and identify molecules that may be at very low concentration. While emphasis will be placed on explosives, it is understood the application fast detection of any other material such as toxins in a food production line. Despite significant progress, dogs are still better than any method implemented to date for detecting explosives. Dogs use their very specialized sense of smell for this task and therefore require a significant number of molecules in the air to detect an explosive. Turns out that most solid or plastic explosives have extremely low vapor pressure. Dogs, therefore, sniff compounds that are associated with the explosives or with compounds that result from degradation of the explosives. Given this explanation, there are two alternatives: detecting explosives or related compounds by sampling the air, or detecting the explosives in their solid-state form. The former approach can use the femtosecond laser as a source of ionization. This works very well; however, advances in the design of robust sources of vacuum ultraviolet light and the use of novel mass spectrometers capable of causing molecular fragmentation for identifying molecules lead to systems that are robust and compact and have a lower cost than a comparable system requiring a femtosecond laser. Therefore, I address here the latter challenge, related to detecting at a distance solid-state microparticles containing trace quantities of explosives.

The greatest advantage of lasers for standoff detection is that the laser can be directed and focused on a target that is tens of meters away. The laser can be scanned over large areas, and if the laser is of sufficiently low power and at a wavelength that does not penetrate the eye, then it is safe to be used in public spaces without restrictions. It is important that the laser can detect a large variety of molecules; therefore, it makes sense to focus on the vibrational signature of molecules for identification. Turns out that one can use spontaneous Raman spectroscopy to detect the vibrational signature of molecules using a conventional laser. Unfortunately, the probability of detecting a Raman-shifted photon with vibrational information is about 1 per 100 million photons. This implies that the laser needs to be quite powerful. Using pulses with very short wavelengths can ameliorate this drawback.

Unfortunately, the UV pulses can cause damage to skin and can even break chemical bonds. Therefore, even when material is detected, the laser being used for detection is deteriorating the target. Given that these alternative methods are not ideal, there is a need to develop an alternative, as described below.

At this point, it is worth recalling the discussion related to CARS imaging for biomedical imaging. The CARS process involves excitation with a broadband femtosecond laser to drive the molecular vibrations coherently. One can imagine at this point a microcrystal where the majority of the molecules are vibrating at a specific modulation rate. That modulation, in turn, modulates a narrow-bandwidth probe laser, thus causing side-bands in the spectrum that are known as the CARS-shifted emission. This is the foundation of standoff CARS for trace explosives detection. Alternatively, the modulation can change the spectrum of the scattered light, indicated by transferring photons from the high-energy side of the spectrum to the low-energy side of the spectrum. This process is known as stimulated Raman scattering (SRS). By measuring that spectral shift, one is able to very sensitively detect the presence of a trace quantity of explosives.

My research group has been working on developing a laser-based approach that could address the challenge of trace explosives detection. We worked on ionization in the early days, because at that time, the initial goal of the project was for detecting a chemical threat in the air. Our most impressive accomplishment was to be able to demonstrate that not only could we identify different molecules, we could also identify and quantify mixtures of molecular isomers—molecules with the same molecular weight and constituent atoms but in a different structural arrangement. We accomplished the differentiation of molecular isomers by realizing that the laser–molecule interaction was very sensitive to molecular structure. Therefore, by using two different pulses, for example, one transform limited and one shaped, we were able to detect the ratio between different ion yields in the mass spectrometer to differentiate the compounds. While this accomplishment was quite significant, implementation would be expensive and at the time of our first publication, this approach did not represent a viable industrial solution.

We continued our research on a method to detect trace quantities of solid explosives using a laser without the need for additional contrast agents or chemicals. Our approach focused first on the use of the CARS method described above. For this project, however, we needed to generate high-intensity sub–10-fs pulses. In those days, there was only one reliable way to accomplish this. One needed a stable amplified titanium–sapphire femtosecond laser and then a capillary tube filled with argon gas was used to generate a coherent broadband continuum. Upon compression of the continuum, we were able to detect pulses as short as 4.8 fs. We also needed to take a small portion of the laser output and convert it into a narrow band picosecond probe. Using a single pulse shaper allowed us to make the first demonstrations of CARS spectra obtained at a distance of 12 m in a single laser shot. However, in those days, the samples were neat liquids, something much simpler to detect than trace quantities of powder on a scattering surface like a metal. An upgrade of our system, where we used a double pulse shaper, allowed us to generate a much narrower probe and to regulate the intensity of the probe independent

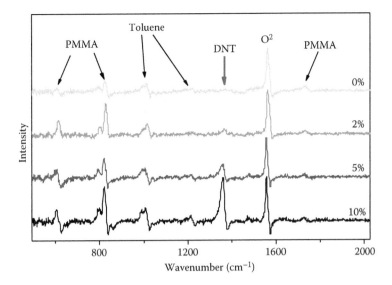

FIGURE 12.1   Detection of trace quantities via CARS spectroscopy of dinitrotoluene (DNT) in polymethyl methacrylate (PMMA) films at a distance of 1 m.

of the broadband pulse. This gave us a very significant increase in signal-to-noise ratios and sensitivity, and demonstrated the presence of trace quantities of nitro-containing explosives mixed with other compounds (Figure 12.1).

Once we demonstrated the concept for our CARS approach to standoff detection, we learned that the ability to scan the laser and to produce an image was of great interest. We therefore proceeded to use shaped laser pulses in order to selectively excite certain vibrational resonances that are common to all explosives with nitro functional groups. Using such a setup, we were able to use a

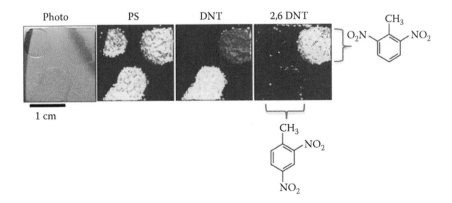

FIGURE 12.2   Trace chemical imaging of two different isomers of dinitrotoluene (DNT) in polystyrene (PS) films at a distance of 1 m.

single photodiode for signal collection, and this sped up the acquisition, making it possible to obtain images in a few seconds, limited only by the 1000 pulses per second repetition rate of our laser (Figure 12.2).

The next research challenge was to test if our approach would work on surfaces such as wood, clothing, or plastics. We opted to test the use of SRS approach, as described earlier. In SRS, the signal is just scattered light. When combined with selective vibrational mode excitation, we built a system capable of determining whether energy transfer associated with SRS had taken place. During SRS, energy from the pulse is transferred from the bluer wavelengths to the redder wavelengths. We configured a pair of detectors to sense differences in the spectral weight of the two wavelength ranges with great sensitivity. This allowed us to detect trace quantities (sub-microgram) per centimeter squared on a wide variety of surfaces including plastic, wood, and clothing (Figure 12.3). As a reference, a grain of salt weighs approximately 50 mg, about 10,000 times more than the amount our system can detect.

More recently we designed a new system that is able to accomplish the same level of sensitivity but with greater speed and using an eye-safe wavelength laser. Our approach is based on using an industrial erbium-doped fiber laser and amplifier and a photonic crystal fiber rod to generate the broad bandwidth. A pulse shaper compresses the pulses and can be used for selective vibrational state excitation. This system has been used to detect a variety of compounds including explosives in a variety of surfaces including car panels and car windshields (Figure 12.4). The laser system being used for this approach operates at a repetition rate that is 2000 times faster than our previous laser. This allows us to acquire multiple chemically resolved images in less than 1 s.

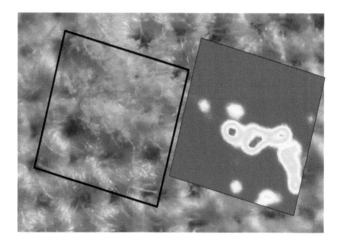

FIGURE 12.3   Trace chemical imaging of the explosive compound known as RDX on a cotton fabric. The image was obtained by SRS at a distance of 7 m.

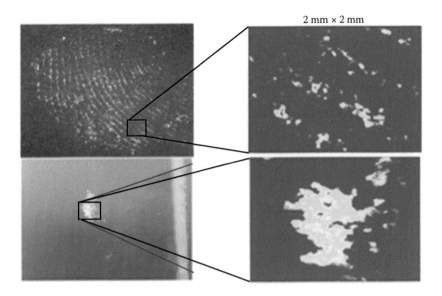

FIGURE 12.4   Trace chemical imaging of sulfur powder (top) on a red car panel and of triacetone triperoxide (TATP) (bottom) through a 6-mm car wind shield. The images were obtained by an eye and skin safe laser at a distance of 0.2 m.

In summary, femtosecond lasers, because of their broad bandwidth, make vibrational spectroscopies such as CARS and SRS practical and efficient. In this chapter, I described how these lasers have been used for the identification of trace quantities of explosives. The reader can imagine that the same approach can be used to detect traces of multiple different compounds in different contexts.

13 Applications of Shaped Pulses to Surgery and Material Cutting or Processing

Presently, a laser with sufficient intensity to cut optical tissue is focused directly into the eye in order to improve human vision. This procedure is not science fiction; it is one that is performed thousands of times per day worldwide. It is one of the most successful industrial applications of femtosecond laser surgery and material processing. The development of this technology required femtosecond lasers and pulse shaping, at least to the extent required to deliver reproducibly femtosecond pulses to the cornea. The history of this development can be traced back to experiments in the 1950s involving surgical cuts designed to reshape the cornea, and some of the earliest ruby-laser experiments, in particular those involving the observation of optical breakdown in water in the late 1960s. What was surprising about laser breakdown was that water is a transparent liquid without absorption at the wavelength of the laser. The breakdown process was occurring through a nonlinear optical absorption, processes discussed in Chapter 6. By the 1980s, the neodymium pulsed yttrium aluminum garnet (Nd-YAG) laser had been developed, and optical breakdown was beginning to be tested for disrupting optical tissues. The quality of the early experiments benefited from a method called femtosecond laser pulse optical ranging that was the precursor to optical coherence tomography (OCT). It was found that the bubbles resulting from cavitation after optical breakdown was causing significant collateral damage to the tissue. During this time, an alternate technique for eye vision correction using excimer lasers at 193 nm was developed. In that case, the very short wavelength laser light was absorbed by the first few micrometers of the corneal tissue and was thus ablated with great precisions. Therefore, in the 1990s, excimer lasers dominated visual correction surgeries until the early 2000s, through a procedure that became known as laser-assisted in situ keratomileusis or LASIK. During LASIK, a microtome blade is used to cut a flap out of the upper layer of the cornea. The flap is then folded back to expose the inner part of the cornea. The excimer laser is then used to remodel the cornea to correct the focusing properties of the eye. Following remodeling, the flap is returned to its place where it heals in place.

Reducing the collateral damage caused by laser-induced breakdown greatly benefited from research in the late 1980s comparing near threshold corneal ablation using nano-, pico-, and femtosecond lasers. It was found that pico- and femto-second laser pulses near threshold could approach the same level of precision as the excimer laser. By mid-1990s, research on femtosecond ablation of human corneas got started in earnest. However, these efforts had an academic research perspective. The development of femtosecond laser systems for eyesight correction started with an accident at a laboratory at the University of Michigan. One of the scientists working on a high-power femtosecond laser inadvertently got a laser focused on his eye. He was rushed to the eye health center on campus and the doctor that examined him could not believe the damage that the laser had caused. The eye had a perfectly clean cut near the surface of the eye. The cut did not seem to have an entry or exit wound. Furthermore, there was absolutely no damage to the retina, the portion of the eye that contains all the photoreceptors. Therefore, the scientist had an excellent recovery. The eye doctor was very excited about this occurrence. He had witnessed the perfect eye surgery tool in that accident and started thinking about its development. Fortuitously, at the same university, there were experts in femtosecond laser design and experts in the development of high-repetition rate picosecond Nd-YAG lasers. After some proof-of-concept experiments, the newly formed team set out to develop an automated system for assisting photorefractive correction taking advantage of the high-quality ablation occurring when near-threshold (3×10^{12} W/cm^2) sub-picosecond laser pulses are focused on transparent tissues. During the early human trials, it was found that sub-picosecond pulse duration was required to get optimal outcomes. This required spectral phase dispersion management, which for these pulses could be accomplished by grating pairs. In addition, it was found that the femtosecond laser could be used to create the flap that is used during LASIK to remove the upper surface of the cornea before excimer ablation (Figure 13.1). Given that most of the complications with the earlier LASIK procedures stemmed from the flap, there was tremendous potential for improvements. The creation of the femtosecond LASIK flap was aided by the use of a flat surface that flattened the eye and thus permitted highly precise flap creation.

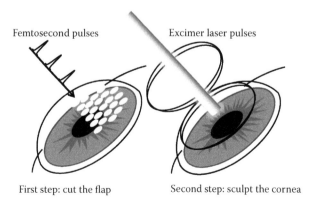

FIGURE 13.1   Femtosecond laser eye correction surgery.

Further advances in the use of femtosecond lasers for ophthalmology have included the development of higher-repetition rate lasers (going from hundreds of kilohertz to megahertz), lower-energy-per-pulse lasers (from microjoule to nanojoule), and shorter-pulse lasers (from up to 800 fs to 250 fs). The progression toward shorter, less energetic pulses has been shown to further reduce collateral damage and to produce much smoother tissue interfaces. It can be expected that this trend will continue as robust femtosecond lasers capable of producing shorter laser pulses become available. These efforts may involve the development of fiber lasers capable of directly generating nanojoule, sub–50-fs pulses at up to 100-MHz repetition rates. Once again, pulse shaping will be required to ensure consistent delivery of short pulses to the focal plane of the laser. Given the involvement of fiber lasers and microscope objectives, it is likely that third-, fourth-, and higher-order dispersion will need to be corrected to achieve clean short pulses. Pulse shaping may enter the implementation of future femtosecond ophthalmic systems (a) through laser pulse characterization for the tailored design of pulse compressors, (b) through laser pulse sculpting for improvements in efficiency and efficacy, and (c) for the creation of versatile systems capable of delivering different output pulses depending on the specific medical procedure being performed.

Success in refractive surgery inspired ophthalmologists to address the need for skilled surgeons to perform cataract surgery. Cataract surgery involves the removal of the lens in the eye and its replacement by an artificial lens. As we all age, the lens inside our eyes loses its transparency. Therefore, if we live long enough, we are likely to need cataract surgery. For this procedure, the femtosecond laser is used to make all the necessary incisions to remove the lens of the eye. Success in this type of surgery requires accurate centering of the artificial lens. The femtosecond cataract system is guided by a computer and is able to make excellent incisions without the need of an expert. While skilled surgeons are able to carry out highly successful cataract surgeries, there aren't enough skilled surgeons to address the need for cataract surgeries for the aging population in the world.

Industrial laser cutting, as being used at present, and what we could call first-generation laser processing, involves laser light that melts a small region in the metal and uses compressed air to blow the molten material out of the way. Under best conditions, the region that is molten by the laser is from 1 to 3 mm in thickness and can extend to several millimeters in depth. The process is one of focused melting, and the speed of the cut depends on how much energy the laser can deliver. The development of ytterbium fiber lasers producing tens of kilowatts has made fast laser cutting cost-effective and is now replacing the older-generation CO_2 lasers. Note that first-generation laser processing described in this paragraph does not involve short pulses. It is a process of sheer energy delivery where peak intensity is not of importance. The need for femtosecond pulses for material processing, cutting, and dicing arises when one is interested in cuts that are very fine (measured in microns) and one cannot accept a scar on the cut created by melting. Moreover, first-generation laser processing cannot work on transparent materials because the light just transmits through instead of cut the material.

In Chapter 6, we discussed nonlinear laser–matter interactions as a function of laser intensity. We established that at intensities of 10^{13} W/cm^2, the laser ionizes matter. In liquids and solids, avalanche ionization takes place, involving a cascading process of generation, acceleration, high-energy electron atom collisions releasing more electrons, and further acceleration. As a result, the atoms that interacted with the laser (within the focal volume as defined by the focal spot and Rayleigh length) are ablated from the material. Most importantly, the nearby atoms are not affected; in fact, under ideal conditions, the surrounding material does not melt. This process, also known in the industry as cold ablation, is very different from first generation industrial cutting with powerful lasers having long pulses or not being pulsed at all.

With pulse durations of 10^{-13} seconds, pulses with 1 μJ of energy have intensities of 10^7 W, and can reach peak intensities of 10^{13} W/cm^2 when focused to a 10-μm spot size. At these intensities, nonlinear optical processes dominate. Therefore, instead of simply transmitting through a transparent material, the laser ablates the material. The ablation process proceeds through ionization and Coulomb explosion rather than the more conventional melting followed by evaporation. The process occurs only within the volume where the laser is focused. With high numerical aperture objectives, it is possible to restrict the focal volume to 1 μm in three dimensions.

Realizing that the femtosecond laser only ablates the material that is at the focal plane of the laser has inspired scientists and engineers to develop systems that scan the focal plane of the laser over materials that need to be polished. Maintaining the peak intensity low ensures that only features that protrude through the surface being polished get ablated.

In the world of semiconductor chips, microprocessors, memory chips, integrated circuits, and even light-emitting diodes, a key step in their production is to cut these devices from the wafer. A wafer, a disk that can be as large as 0.450 m in diameter, can contain hundreds of chips. The chips, in the early days, used to be cut out using special diamond disks. Newer methods included scribing with a diamond-cutting tool and then cleaving. Femtosecond lasers are now being used for scribing these components. Femtosecond laser scribing has shown to leave a minimal amount of debris, cause a minimal loss of material in between the chips, and most importantly, in layered materials, the femtosecond scribe followed by cleavage, there is minimal interlayer contamination, which has led to much higher efficiency light-emitting diodes.

One of the latest industrial applications of femtosecond lasers is the correction of individual pixels in large area organic light-emitting diode screens. The newer monitors, with millions of pixels, can have a few defects. While efforts are underway to reduce the number of defects to zero, femtosecond lasers have been shown to be capable of correcting these individual pixels. The proposed method would have an imaging system that locates the pixels that need to be fixed by the laser; the laser is then aimed and the laser zaps the pixel. The fully automated system could be incorporated into the production line. At the time of the writing of this book, femtosecond lasers with pulse durations shorter than 500 fs are being evaluated for this task.

Further industrial applications of femtosecond lasers will take advantage of the high degree of localization of the ablated spot. In terms of energy consumption, if the laser ablates one-hundredth of the material that is presently being ablated by first- and second-generation laser material processing, then significant energy savings can be realized (Figure 13.2). For example, assuming the process of cutting steel presently melts 10,000 mm^3 of metal per meter, the energy savings would be enormous if that number could be reduced to 100 mm^3 by using ultrafast lasers. Moreover, there would be significant materials savings, and the quality of the cuts would be greatly superior. This implies that post-cut processing may be completely eliminated. Another exciting application of femtosecond material processing comes from the ability to process composite and layered materials. Once again, the ability to localize the ablation process with micrometer accuracy permits the processing of metals, polymers, and ceramics in layered composites. Processing such materials is difficult because each material has specific properties. For example, conductive metals are easily cut through electric discharge machining (EDM); however, when they are protected by a ceramic thermal barrier as in turbine blades, they can no longer be drilled by EDM. The femtosecond laser can be tailored to optimally cut one or all of the different materials. As the reader may surmise, material processing with femtosecond lasers is not as simple as exceeding the ablation threshold. When the intense pulse interacts with the material, the plasma formed in the initial stages of the interaction could reflect a larger percent of the pulse energy if the pulses are very intense. Evidence of this process comes from reflectance measurements, as well as from experiments in which an intense laser was split into multiple pulses. It has been found that the rate of material removal can be improved by factors greater than 50 times through splitting the femtosecond pulse into multiple pulses. The concept behind this significant improvements is that the pulses should only be intense enough to remove material but at the same time to minimize plasma reflectance. By splitting the pulse, one is able to increase the repetition rate of the ablation process while keeping the efficiency as optimum as possible.

So far, most of the applications being considered have been related to ablation. The femtosecond laser can also be used to modify the index of refraction of transparent materials. Within transparent materials, one is able to trace a circuit, in a sense writing a waveguide for light. This approach has been demonstrated and has

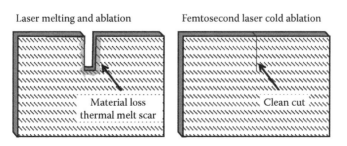

FIGURE 13.2   First-generation versus third-generation femtosecond laser cutting.

been used for some applications. The ability to sculpt the index of refraction of materials can be used for writing gratings in large volume glass or within optical fibers. These gratings can be used as filters or as means to correct dispersion of pulses being transmitted through the fibers. It is conceivable, that, in the future, femtosecond lasers will be used to tailor optics that will in turn be used to manufacture superior femtosecond lasers. Finally, through manipulation of the index of refraction of the human cornea, it is possible to improve vision without the need to ablate material as described earlier in this chapter. This procedure is presently being evaluated for commercialization.

While the femtosecond laser is ablating the surface of a metal, the electron waves known as plasmons, formed in the surface interfere. The plasmon interference causes the surface, instead of being polished, to develop a structured pattern that resembles lines with nanometer thickness and separation. Because these features are smaller than the wavelength of light, they absorb all wavelengths or they can have unusual surface properties such as being extremely water repellant or water absorbent. At present, these surface processing methods are very expensive for large area applications. However, as femtosecond lasers become industrialized (see Chapter 17), they are sure to become cost-effective. It seems the first commercially viable applications for femtosecond laser surface patterning are direct writing on jewelry, or the creation of semiconductor sensors with superior sensitivity.

14 Applications of Shaped Pulses to Communications

In this chapter, I address optical communications, in particular optical fiber based–communications. Among the key components required for a communication system, we find (a) short-pulse oscillator, (b) data modulator, (c) optical fiber, (d) signal amplifier, and (e) receiver (Figure 14.1). Given the tremendous need for communication systems in the world, each of these subsystems needs to be highly scalable and robust. Optical fiber–based communication was developed in the 1970s with the successful development of optical fibers and the GaAs semiconductor laser. Digital data transfer rates are measured in bits per second, where each bit is represented by a laser pulse (1) or the lack of a laser pulse (0). The standard prefixes apply, so 1 Mb/s corresponds to a million, 1 Gb/s to a billion, 1 Tb/s to a trillion, and 1 Pb/s to a quadrillion bits per second. Note that 10 Tb/s already implies a time between bits of information better than 100 fs. While first generation systems worked at <10 Mb/s, second-generation systems based on 1.3 μm InGaAsP semiconductor lasers achieved bit rates of up to 1.7 Gb/s with a repeater every 50 km. Third-generation systems developed in the 1980s, operating at 1.55 μm, led to 2.5 Gb/s with the newly developed erbium-doped fiber-laser amplifier replacing repeaters, with a spacing in excess of 100 km. Fourth-generation fiber optic systems operational by the year 2000 achieved bit rates in the 10 Tb/s with an amplifier spacing of 160 km. While these transfer rates seem impressive, Google has indicated that their servers collectively are able to transmit 1 Pb/s. This is not the computer-to-computer data transfer rate, this is the collective capability of their data center. However, the fast growth of the Internet is placing a high demand on ever faster data transfer rates. For example, the transmission of 5000 HDTV 2-h-long videos in 1 s requires 1 Pb/s data transmission rate. Here, I will discuss how ultrafast laser pulses and pulse shaping have played and will continue to play a key role in these developments.

We start by realizing that the ability to read data being transmitted through an optical fiber requires the symbols or pulses to remain distinguishable. If the pulses get broadened while being transmitted, they can start to blend in and this affects their readability (Figure 14.2). The key source for pulse broadening is spectral dispersion as discussed in Chapters 4 and 5. Broadening is proportional to the inverse pulse duration squared. That implies that for relatively long pulses, for example, half a nanosecond, dispersion causes a million times less broadening than for half a picosecond pulse. This is important because higher transmission data rates are very desirable. For example, 1 Gb/s data transfer rate implies 1 bit is transmitted with

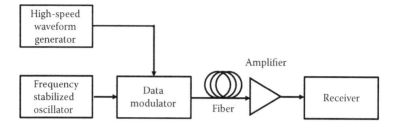

FIGURE 14.1   Optical fiber communication system (schematic).

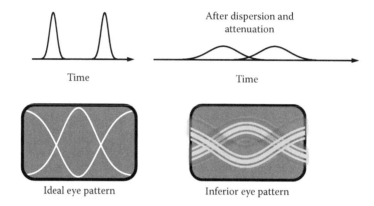

FIGURE 14.2   Signal degradation in optical fiber communication.

subnanosecond time, and 1 Tb/s requires each bit to be transmitted with subpicosecond time. However, when the pulses being transmitted exceed the bandwidth of the fiber bandwidth, the communicated signal becomes distorted because of temporal overlap and interference between pulses. Optical communication systems are evaluated using the "eye pattern" measurement, which is an oscilloscope trace showing the detected signal as a function of the transmission rate. When the signals come in cleanly, an open eye pattern is observed. As the interpulse interference increases because of jitter, or dispersion, the eye pattern becomes fussier and the open regions decrease. The transition from Gb/s to Tb/s required a number of important developments in ultrafast lasers and pulse shaping.

The evolution of modern optical communication systems can be traced to key developments at Bell Laboratories where the first femtosecond lasers (dye based and fiber based) were developed in the mid-1970s. It was immediately realized, as shorter-pulsed lasers were designed, that dispersion compensation was critical. Moreover, it was realized that optical fibers introduced higher-order dispersion (Chapters 4 and 5), and when the input pulses were very intense, additional pulse distortions arising from self-action further distorted the pulses. Being able to mitigate all these pulse distortions would require systems capable of manipulating the pulses well beyond what simple static optics could accomplish. Hence, the efforts into pulse shaping begun in the early 1970s. At first, these efforts took advantage of grating pair compressors. In the early 1980s, the 4f pulse shaper was introduced, and by the late 1980s, it was already being used in laboratories to manipulate the spectral phase of pulses, however, using fixed phase and amplitude at the Fourier plane. The programmable liquid crystal spatial light modulator to manipulate phase and then phase and amplitude became available in 1990. During this time, it was realized that the phase and amplitude of a highly chirped pulse could be modulated with an electro-optic modulator. An acousto-optic tunable filter was then introduced to modulate the amplitude of multiple frequencies in the early 1990s, and by 2000, this approach was incorporated into a single optic.

The systems being developed in research laboratories by the year 2000, using very short-pulsed lasers, sophisticated pulse shapers, and ultrafast detectors, were far from scalable implementation. Exceeding 10 Tb/s data rates, therefore, had to wait for these technologies to mature. Fortunately, tremendous progress could be done by implementing concepts that came from radar and radio communications while taking advantage of existing optical components. This latter alternative led to developments that immediately increased data transmission bandwidths, and many of these approaches now take advantage of improvements in all the optical components. In particular, I introduce here six highly relevant approaches to modern optical communications; the first three are illustrated in Figure 14.3.

Time division multiplexing (TDM): Is an approach in which a single data string aggregates information from multiple sources. This approach conceptually dates back to the 1950s and requires the transmission medium

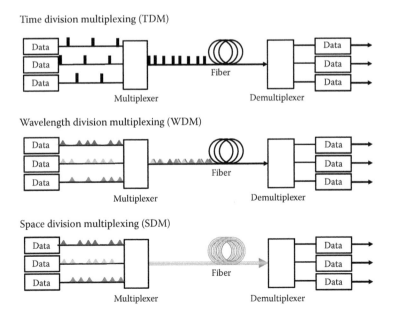

FIGURE 14.3 Multiplexing approaches to optical fiber communications.

to have a bandwidth that exceeds the aggregate bandwidth of all sources in tandem. Alternatively, instead of multiple sources, a pulse shaper can be used to encode additional information into each pulse, thus sending a byte of data every pulse. Using TDM, data transmission of 100 Gb/s were demonstrated in the 1990s.

Wavelength division multiplexing (WDM): It takes advantage of spectral bandwidth in optical fibers by separating the information into independent wavelength channels. In this approach, the temporal requirements for each wavelength channel are relaxed. In a sense, each wavelength channel performs at a rate that can be sustained by standard optical elements. The aggregate of all wavelength channels results in very high data transmission rates. Conceived for optical communication in 1980, this approach has been very successful, and in optical fiber communications, it can presently aggregate hundreds of Gb/s to achieve data transmission at rates of tens of Tb/s. For WDM, pulse shapers can be used to modulate each of the wavelength channels.

Space division multiplexing (SDM): Similar to WDM, in SDM, signals are separated in space rather than in wavelength. This approach has gained popularity owing the recent development of multiple-core optical fibers. Given the spatial separation of signals, this approach requires spatial pulse shaping.

Optical code division multiple access (OCDMA): This approach combines TDM and WDM concepts. First developed in the 1950s for radio communications, this approach has been the backbone of cellular communications. Its implementation in optical fiber communications requires a transmitter consisting of a short laser pulse oscillator being modulated by a data modulator. The pulse shaper then encodes an address, breaking the data pulses into multiple pulses such as in TDM. The receiver uses a pulse shaper with a conjugate phase mask to identify the message and decode the information. Incorrectly addressed signals are rejected. This approach is promising but it requires dispersion compensation because of the use of sub-picosecond pulses.

Polarization division multiplexed M-quadrature amplitude modulation (PDM-MQAM): In this approach, two orthogonal polarization states are used to send M-bit quadrature amplitude modulated signals. Each symbol carries M/2 bits of information, resulting in 16, 32, and 64 data transmissions. This approach combines TDM, WDM, and PDM.

Electrical dispersion compensation (EDC): This approach is not a form of multiplexing; it is rather the development of a computer-based pulse shaper designed to retrieve signals that have suffered dispersion. This approach can be conceptualized as one of signal recovery through deconvolution, a process in which a pair of pulses that have been blurred are re-sharpened after the signal has been detected through digital signal processing. If two pulses are blurred, standard deconvolution approaches would not work because the information has been lost. In these cases, it is important to understand the different sources of pulse degradation (dispersion, self-phase modulation ([SPM]), and interference) and to develop an "intelligent" system that learns how to distinguish the possible different input signals. This last step is known as a maximum likelihood sequence estimator. Typical improvements based on this approach can extend the dispersion length of fiber by factors of 5 to 10. Modern EDC chips can perform adaptively, taking into account time-dependent degradation in the optical signal transmission lines.

Having described the different multiplexing approaches, the reader can quickly realize that all these approaches can be used in tandem. For example, each core in an SDM system can carry multiple wavelengths taking advantage of WDM, and each wavelength can be further encoded by TDM. Now, I return to the developments that have brought optical communications to their present rates, and how we may continue to maintain the tremendous rate of progress. The implementation of WDM required optical components capable of wavelength routing of the different data streams. Such a device was developed in the mid-1990s. ODCMA was demonstrated using a pulse shaper in the late 1990s over a 2.5-km distance.

At this point, it is worth considering the optical fibers themselves. Optical fibers are quite remarkable because they transmit laser light as any other type of cable, without the requisite of traveling in a straight line. What is more remarkable is that optical fibers are highly transparent. For example, a 10-km (equivalent to 6 mi) fiber made out of ZBLAN glass (with a composition ZrF4-BaF2-LaF3-AlF3-NaF) transmits 80% of the input light. This is approximately the transparency of a car windshield. The development of single-mode ZBLAN optical fibers with dispersion shifted to 1.55 μm resulted in 100-km waveguides supporting 10-Gb/s data transmission rates. The waveguide dispersion can be tailored through the addition of different salts (as in ZBLAN). In addition, the dispersion can be further fine-tuned by manipulating the core diameter and the refractive index of the cladding. Dispersion manipulation in fibers has led to the development of fibers that have the opposite dispersion sign at a given wavelength. This helps eliminate second-order dispersion; however, it does not correct for third- and higher-order dispersion.

The effects of spectral dispersion can be addressed primarily by using sections of fiber with opposite dispersion. However, long-distance transmission requires higher input pulse energy. Typically, pulse energies are of the order of 0.1 nJ. At higher pulse energies, SPM degrades the pulse recompression process and provides an upper boundary on the transmitted pulse energy (see Chapter 6). Given that SPM depends on peak intensity, it plays a more important role when the pulse is fully compensated (compressed) and has the shortest duration. In principle, it is best if the pulses are not compressed during the majority of the transmission distance.

A number of communication-specific pulse shapers have been described in the scientific literature. Beyond the acousto-optic modulators and liquid-crystal spatial light modulators, one finds descriptions of micro-optomechanical modulators and other micro-electro-mechanical devices designed to take the wavelength-dispersed input spectrum and modulate each wavelength and send the information taking advantage of WDM and OCDMA. These types of devices can also be used for decoding highly multiplexed information. The cost and complexity of modulators have also inspired the design of optics that passively accomplish encoding; an example of this type of pulse shaper is the distributed Bragg grating. All these developments depend on faster modulators. Gigahertz modulators are already deployed in communications and there are a significant number of proposed terahertz modulators.

Another exciting development that may shape the future of communication systems is the digitally controlled chirped laser (DCCL). The output of this type of laser is naturally chirped. This makes it naturally pre-compensated for long-haul communications. Alternatively, because its frequencies are being separated in the time domain, it can be shaped with fast modulators before transmission. DCCLs are being miniaturized and their chirp can be controlled electronically.

Taking advantage of the first two multiplexing methods illustrated in Figure 14.3, data rates of 100 Tb/s were demonstrated over 7-km distances in 2010. At the time this book was written, 32-core fibers have already been introduced. Using these new fibers with SDM over 32 cores and PDM-16QAM has led to the demonstration of transmission speeds of 1 Pb/s over 200 km, a result reported in 2017. Further improvements are expected once a larger spectral bandwidth is used. Such a development requires a second amplifier for the additional bandwidth. Note that this impressive achievement takes advantage of multilevel spectral modulation, multicore waveguides, and multimode control including PDM. The multiplicative scaling of the approaches mentioned bodes well for continued improvements in data communication rates that will keep up with Moore's Law for the foreseeable future.

Pulse shaping will continue to be an integral part of optics-based communications. Here, I briefly discuss some of the present challenges. For long-haul communication, there is a need to transmit high-energy pulses for hundreds to thousands of kilometers without the need for repeaters. It is likely that optical satellite links will help the longest-distance communication, given the much lower dispersion of air compared to optical fibers. For 100- to 1000-km-distances, optical-fiber communication, combinations of the multiplexing methods mentioned above should be sufficient to reach hundreds of Pb/s. As wireless communication continues to evolve, there may be a need to shift toward the visible spectrum. In this case, light will be modulated at gigahertz rates and detectors will need to retrieve information after multiple scatterings. Finally, the highest demand will be for relatively short-distance communication between servers. The aggregated data rate of such systems places the highest requirements. Discussion for those optical systems is moving toward hundreds of terahertz, equivalent to 10-fs pulses. Given the information in Chapters 4 through 6, it is likely that these systems will not involve a single core. However, even with multiple cores, further improvements will require ultrafast detectors, ultrafast optical switches, and advanced pulse shaping that involves space and time distribution and aggregation of signals. Systems will require adaptive dispersion characterization and compensation, the reader can find more ideas regarding future developments in Chapter 17. In addition, new fibers will be needed. In this vein, three developments stand out: hollow core fibers that offer minimal dispersion, photonic-crystal fibers with tailored dispersion, and polymer fibers with fully tailored dispersion.

15 Applications of Shaped Pulses in Science

I described in the prologue how an octave spanning femtosecond laser can generate any electromagnetic frequency through wave-mixing or multiphotonic processes. The scientific applications of such a universal light source depend on one's ability to control the wave-mixing or multiphotonic process. That control is achieved through spectral phase manipulation with a pulse shaper. Pulse shaping was in the minds of scientists well before femtosecond lasers were demonstrated. Initial research involved simple pulse compression. During those early days, nonlinear optical processes, such as second harmonic generation, were being studied for fundamental curiosity and also because they could be considered for optical switching. Therefore, scientists were exploring how pulse shaping affected second harmonic generation. One of the interesting findings was that a highly dispersed pulse could be made to generate much higher second harmonic signal by blocking some portions of the spectrum using a pulse shaper. This finding seemed to defy intuition because it appeared that attenuating the input light by blocking some of its spectrum increased the intensity of the output second harmonic signal (Figure 15.1). The explanation for this finding was that some frequencies seemed to be interfering with others and that blocking the interfering frequencies caused the observed higher second harmonic signal. In the example included here (Figure 15.1), Gaussian pulses (dashed line) stretched 50 times their original duration by a quadratic phase (black). The peak intensity of the pulse and SHG intensity induced by this pulse drops by a factor of 50 times because of the longer pulse duration. When the spectral components that are out of phase are blocked by setting their amplitude to zero, the pulses are compressed and the SHG intensity recovers to about 15% that of transform-limited pulses. The lower row presents results for binary-phase compressed pulses, where the out-of-phase components are rephased by applying a pi step. This time, the pulse compression is much better and a peak intensity of ~0.4 is achieved. Notice the time profiles of the TL pulse (dashed) and the binary compressed pulses (red), which are very close to one another, with pulse and duration of the binary compressed pulse practically equal to the duration of the TL pulse. From these figures, we learn that phase compression is more efficient than amplitude modulation because one is able to turn destructive interference into constructive interference.

Early research into pulse shaping was aimed at developing an optical arbitrary waveform generator (OAWG). Such a device would be able to create any type of pulse, for example, square pulses or triangular pulses of any duration, at will, like an arbitrary function generator in electronics. An OAWG manipulates each frequency within the bandwidth of the pulse, and the greater the bandwidth,

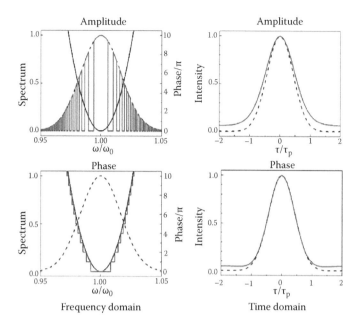

FIGURE 15.1   Amplitude (top row) versus phase (bottom row) pulse compression to maximize SHG intensity.

the greater the capabilities of the system. The OAWG can be used to design pulse trains (Figure 15.2), or pulses with slow rising or slow decaying fields. In a sense, the OAWG, when properly calibrated, delivers precisely tailored pulses that can be confirmed by a number of pulse characterization methods. Already, some advanced attosecond laser sources include a pulse shaper running MIIPS, discussed in Chapter 10, to ensure that the high-harmonic generation process is always optimized. The pulse shaper can determine the precise delay time and phase difference between the two pulses. A well-designed and well-built pulse

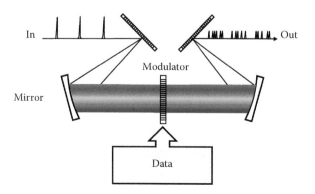

FIGURE 15.2   Conceptual diagram of an optical arbitrary waveform generator (OAWG).

shaper, being a common path interferometer, provides outstanding stability. It is easy to confirm that the shaper can step the time delay between pulses with 10 attosecond accuracy. Some groups have claimed sub-attosecond stability. As of the writing of this book, OAWG systems operating with GHz modulators are already being used to encode communication data.

The use of shaped laser pulses for controlling molecules and their chemical reactions was proposed in the 1990s and inspired a number of chemists and physicists. Two different philosophies have been followed (Figure 15.3). The first focused on manipulating the timing between two or more different laser pulses in order to lead a chemical reaction dynamically from beginning to end. In this approach, each pulse causes a transition from one state to a different state. A typical example, known as "pump-dump," involves a first laser pulse to excite a molecule to an excited state from which a given bond starts to stretch. A second pulse then drives the excited molecular state back to the ground state, at a point that ensures breaking of the chosen chemical bond. Given the very different wavelengths needed for accomplishing these multipulse experiments, the majority have been carried out using separate laser pulses rather than a single pulse shaper.

The second philosophy considered the shaped pulses as being able to explore a nearly infinite number of phases and amplitudes of light that upon interaction with molecules would be capable of directing a chemical reaction. Given that a femtosecond laser is intense enough to cause rotational, vibrational, and electronic excitation, via linear and nonlinear optical processes it was proposed that a shaped femtosecond pulse would be able to manipulate the molecule from reagent to the desired reactant. Details of the necessary control that would be required were not computable; therefore, the proposal was to use a pulse shaper to explore any amplitude and phase for all the frequencies of the femtosecond laser. In theory, without restriction of amplitudes and phases or number of frequencies being explored, there seemed to be a 100% probability that shaped lasers could control chemical reactions with perfect yield. The experimental approach to be followed

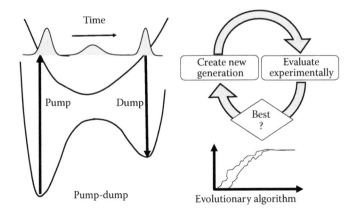

FIGURE 15.3 Conceptual approaches to controlling chemical reactions.

was to take advantage of evolutionary (genetic) algorithms that would evolve a solution from an essentially random starting laser pulse. In a sense, outcomes from the laser–molecule interaction would guide the learning algorithm and it would evolve towards a solution for the problem.

The reality of the genetic algorithm approach was quite different from the theoretical proposal. In those days, the bandwidth available from commercial femtosecond pulses was about 1/20th of an octave. A device having approximately 100 pixels then shaped the laser, and the resolution of the shaper was 10-bit resolution, allowing for 1000 different values of phase and amplitude. Still, such a pulse shaper was able to generate approximately 10^{600} different pulses ($[1000^{100}]^2 = [10^{300}]^2 = 10^{600}$). The lasers being used had a repetition rate of 1 kHz, and because of noise, no more than 100 different pulse shapes could be tested per second. At this rate, fewer than 10^7 pulses could be tested in one day. The number of shaped pulses needed to explore a space that was hundreds of orders of magnitude greater was so large that it could never be explored, not even in billions of years.

What was most surprising in the early experiments using genetic algorithms was that they converged in a matter of a few hours to a solution that could not be improved upon after restarting the algorithm from different random sets of pulses. The first experiments were aimed at pulse compression. The goal was to evaluate different randomly chosen phases in an effort to find one that mitigated phase distortions and resulted in shorter pulses. Pulse duration was measured by detecting the SHG signal generated by the shaped laser pulse. As the algorithm progressed, it would choose a few of the best phase functions and then use those to generate new phases to be evaluated. It was found that within 100 generations, one would obtain pulses that were close to transform limited. By adding certain restrictions such as avoiding very fast jumps in phase, and realizing that the majority of the dispersion would be second- and third-order dispersion, these algorithms were made even more efficient. Pulse compression happens to be an ideal case for computer learning algorithms because there is a clear optimum value, corresponding to transform-limited pulses. Furthermore, deviations from transform-limited pulses reduce SHG signal proportionally. If one were able to map the SHG value for all the possible phases, one would find that there is one clear maximum in the landscape, a smooth Gaussian-shaped mountain (see Figure 15.4). In such a landscape, it is easy to take random steps and to distinguish which steps move us to higher ground and which do not. This type of search space is known as "convex," and it is most efficiently navigated using gradient-following algorithms. In a convex space, one should be able to find the maximum value by evaluating an extremely small sample of the entire search space.

Not every problem addressed by a learning algorithm has a convex space. In the prior analogy, one can imagine that the presence of noise would make the terrain bumpier and the possibility of being stuck in a so-called local maximum becomes real. One is able to program steps designed to test if one is at a local or global maximum. There is, however, the opposite of a convex search space. The opposite is known as the "needle in the haystack" search space (Figure 15.4). If the reader imagines an algorithm searching for a needle in the haystack, one

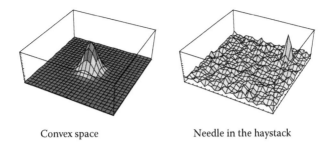

Convex space   Needle in the haystack

FIGURE 15.4 Conceptual possible signal landscapes to be searched by evolutionary algorithms.

clearly realizes that any sparse sampling of the hay does not lead to information that can be used to determine proximity to the needle. In such cases, gradient-following algorithms and learning algorithms are of no avail. In these cases, random sampling and luck seem to be the only methods available. One can, however, greatly increase efficiency by improving speed and by reducing the space that one needs to search.

Reducing the search space can bring about very significant results. Consider the case of the pulse shaper being able to generate 10^{600} pulses. Now, let's start by assuming that the process being optimized is one that involves nonlinear optical interactions with the sample. This assumption stems from the fact that linear optical interactions are best addressed with either incoherent light (very inexpensive) or with laser diodes (fairly inexpensive). If a femtosecond laser is needed, then the interaction is nonlinear. Given that the interaction is nonlinear, then one does not need to use amplitude modulation. This is because one is able to control nonlinear amplitude by using phase. The reader is referred to Chapter 10 where we discussed multiphoton intrapulse interference. Eliminating amplitude modulation results in a dramatic, 300 orders of magnitude reduction in the search space. Reducing the finesse of the phase modulation from 1000 to 100 further reduces the search space by another hundred orders of magnitude, leaving us with a space of size 10^{200}. While this represents a much smaller search space, it is still well beyond evaluation. The conclusion we reach is that unless one is searching a convex space, optimizing a process using genetic algorithms, learning algorithms, or gradient-based algorithms, algorithms offer little or no advantage.

The conclusion above seems to contradict that vast literature reporting the identification of optimum pulses being found using optimization algorithms. I would argue that it reinforces the conclusion when restated as follows. If an optimization algorithm is able to find an optimum solution, the only conclusion is that the optimum solution has a convex search space.

Following on the conclusion reached, it would then seem to be the case that a great majority of the problems that have been addressed scientifically by using optimization algorithms have a convex search space. Here, I offer a simple explanation. We are working with femtosecond pulses, and one can easily consider two extremes. The first would be transform-limited pulses, which would offer the highest peak intensity and would maximize nonlinear optical processes such as SHG, THG, ionization, and saturation of optical transitions. The other extreme would be the longest pulse possible given the experimental constrains of the pulse shaper. Such a long pulse would then minimize nonlinear optical processes. These minimum and maximum optimization extremes were identified early and some groups tracked SHG in order to prevent what they considered to be "trivial solutions." Other groups were not so careful and allowed their shaper to cause all sorts of phase and amplitude modulation without any form of pulse characterization. This has led to results that defy scientific explanation, unless one considers experimental errors.

The appeal of optimization algorithms was that one does not need to know the detailed physics of a given process to optimize it. Fortunately, scientific understanding is quite advanced and we are able to use physics to optimize a process. It takes a millisecond to test if transform-limited pulses or long pulses optimize a process! Similarly, one can test different schemes involving chirp, timing between two pulses, trains of pulses, pulses with a preceding pedestal, or pulses that preferentially excite certain vibrations or excited states in a few seconds. These systematic experiments, known in the literature as "open loop" because they include a thinking scientist, lead to results that can be easily compared with theory and numerical simulations.

Going back to search space reduction, one is able to consider a more drastic step. Imagine that control is about creating constructive interference in the desired path and destructive interference in undesirable pathways. Provided there are no atomic or molecular phases between different excited states, then all one needs is a combination of two phases, zero, and π. This space reduction leads to a search space of 10^{30}, which is much smaller than anything considered so far. Furthermore, these types of binary search spaces have very distinct geometries that lead to fractal search spaces. By reducing the number of active pixels, one is able to get an idea of the search space geometry with 1000 measurements, which take a few seconds. Binary shaping is particularly useful for excitation of sharp transitions either in excited states or in vibrational states, and has been used for standoff detection of explosives. One particular advantage of binary phase–shaped pulses is that they are highly reproducible. Just like in electronics, the use of binary voltages leads to increased robustness; similarly, binary phases give increased robustness of the solutions.

The modern femtosecond laboratory with a pulse shaper can implement an automated pulse compression method such as MIIPS discussed in Chapter 10. It can then use the principles of MII to cause selective two-photon or three-photon excitation to control excitation of different molecules or a given type of molecule to different excited states. Similarly, MII can be used for selective vibrational mode excitation. When applied to biomedical imaging, MII can be used to obtain multicontrast images obtained for differently shaped pulses. Because the pulse shaper does not affect beam pointing, the different excitation modes can be programmed into the pulse shaper without need for realignment. This approach has already been used for multimodal imaging of cancer biopsies in order to better identify the margins of a tumor, as described in Chapter 11. Selective vibrational excitation based on coherence Raman scattering has been used for standoff detection of experiments, as described in Chapter 12.

Some spectroscopic methods known as multidimensional spectroscopy, coherent spectroscopy, and four-wave mixing depend on three or more pulses with well-defined phase between them to excite a molecular system, and sometimes a fourth pulse is mixed with the output signal. The preparation of all these multiple pulses used to involve complicated setups with various optical delay lines. Pulse shapers have started to replace the multiple optical delay lines, providing greater control and stability and control for these types of experiments. One particular type of experiment that takes full advantage of the pulse shaper capabilities is known as phase cycling. Multidimensional spectroscopy experiments may result in signals that are sums of two or more types of phenomena. Separating the signals, for example, the real and imaginary nonlinear optical response of the material, requires one to repeat measurements with different phases between the pulses. The pulse shaper can be easily programmed to change the phase between the pulses by any amount, and the phase cycling process can be fully automated.

Advances in femtosecond lasers have made them more affordable and much more stable than ever before. This has made it possible to consider the use of pulse shapers to tailor the pulses in order to fully control the preparation of specific quantum states. The development of pulse shapers capable of taking pulses in the UV, visible, near-IR, and mid-IR wavelength ranges, including pulses with energies of several millijoules, will enable new types of experiments. The philosophy, in this case, is to control linear (as opposed to multiphotonic) transitions between different quantum mechanical states. The wavelength and pulse duration of the pulses determine the creation of coherent superpositions of states, also known as wave packets. Through phase manipulation, one is able to control, to some extent, the dynamic evolution of the wave packet, particularly its dispersion as a function of time. Multiple such pulses can thus be used to achieve next-generation control of chemical reactions.

The ability of pulse shapers to deliver transform-limited pulses can be used to measure the optical properties of optical media including solids, liquids, and gases. In addition, transform-limited pulses lead to maximum nonlinear optical constructive interference. This makes them extremely sensitive to spectral phase fluctuations. Conversely, the pulse shaper can be used to design destructive interference such that any spectral phase distortion will result in less destructive interference and hence an enhancement in nonlinear optical signal.

Interestingly, the ability of pulse shapers to correct the spectral phase of the pulses in real time will allow laser companies to explore less expensive ultrafast sources. The pulse shaper will be actively correcting the pulses and ensuring the pulses are perfect at all times. In most cases the pulse shapers will be used simply to make sure that transform-limited pulses reach the target reproducibly from day to day, this will bring a new level of accuracy to scientific experiments. In terms of controlling chemistry, advances in laser technology, pulse shaping, quantum

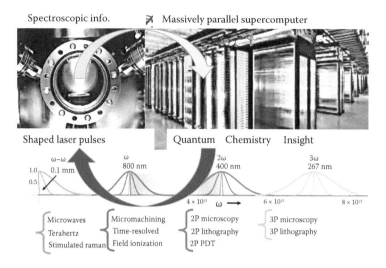

FIGURE 15.5   The concept of a deep-learning spectroscopic system.

theory calculations, and computers will make it possible to explore a chemical reaction through spectroscopic analysis that is informed by high-level quantum chemistry calculations (see Figure 15.5). The experiments are greatly accelerated by the pulse shaper and the high-level quantum chemistry is accelerated by modern computers and the design of modern methods. In conjunction, computer-controlled pulses and high-level quantum chemistry can accomplish the goal of controlling chemistry and, more importantly, of improving important chemical processes.

16 A Survey of New Directions Being Explored, and Potential New Applications

This chapter considers the widespread incorporation of pulse shapers into applications that have already been discussed earlier, such as biomedical imaging (Chapter 11) and standoff detection of explosives (Chapter 12). The emphasis of this chapter is on envisioning how these systems may come together to become valuable integrated systems. I remind the reader that the pulse shaper provides the connection between ultrafast lasers and information technology. The pulse shaper enables full control of the laser pulses for multiple applications. Here, I discuss possible extension of modern ophthalmology tools, where shaped femtosecond laser pulses are used for performing a wide range of procedures. Continuing with surgical procedures in mind, I discuss types of surgeries that could benefit from shaped femtosecond laser pulses. We then consider large-scale femtosecond systems that could be used for the generation of high-energy radiation (x-rays, gamma-rays, electrons, protons, positrons, and neutrons). Such systems could be used for medical procedures as well as for the detection of radioactive materials. I then focus on applications in the field of communications (Chapter 14) to envision systems that could benefit most from shaped femtosecond laser pulses. Finally, I consider new concepts that are beginning to permeate the scientific laboratories that could turn into new future applications.

I discussed the use of shaped laser pulses for the standoff detection of explosives in Chapter 12. Here, I discuss further analytical applications of these laser sources. The ability of femtosecond lasers to excite rotations, vibrations and electronic states makes them ideal for spectroscopic applications. In addition, femtosecond pulses can easily ionize any material. The ions that are formed can then be analyzed by their mass-to-charge ratio for identification. Analytical chemistry, which depends on a wide variety of spectrometers, gives forensic science and drug discovery a solid foundation. Shaped femtosecond laser pulses can be used for carrying out most of the different spectroscopic modalities including gene and protein sequencing. The future of shaped laser pulses in analytical chemistry applications will depend on the development of robust and relatively inexpensive laser sources that include the pulse shaper. These automated sources will then be integrated into multidimensional analytical robots that can be used for widely different tasks such as analyzing a crime scene, for the development of new drugs, for identifying pathogens, and even for space explorations.

I am particularly excited about the use of shaped femtosecond laser pulses for the development of autonomous multimodal imaging systems for diagnostic pathology. The concepts required to make this happen were discussed in Chapter 11. Here, I consider a fully integrated system that is capable of scanning an unstained pathological slide, obtaining chemical information based on multidimensional spectroscopic analysis as discussed in Chapter 11, comparing the findings with databases, and providing possible diagnosis ranked by their likelihood. This development would take advantage of the speed with which all the information can be acquired by the single laser source and the speed of information technology and data processing. The type of pathology workstation being envisioned would provide the function of a pathology department and could be augmented by learning algorithms and databases that feed from numerous hospitals. Even more exciting is the development of these types of pathological systems for the analysis of tissues before they are removed. Dubbed optical biopsy, the concept is to use pathological analysis in living tissue in order to determine if it needs to be excised. The use of such a system in skin is relatively obvious; internal organ analysis such as intestines, esophagus, and bladder can be achieved through the use of endoscopes.

Here, I imagine a short pulse source with less than 10-fs duration that is shaped before entering the microscope. The pulse shaper is programmed to optimize the output and then to achieve optimized multimodal imaging that may include SHG, THG, and TPEF of different compounds; three-photon excitation fluorescence and vibrational excitation at three or more wavelengths; and coherence lifetime imaging. The system will be fully automated and include a number of detectors to simplify data acquisition. The advantage of such systems is that all the technology needed to make them a possibility already exists.

The next possible inroad for pulse shapers is refractive optical surgery, a topic discussed in Chapter 13. Femtosecond laser-based refractive surgery, known as bladeless LASIK, has been carried out with pulses that are between 500 and 990 fs in pulse duration. It has been argued that the precise pulse duration of the laser is of no consequence; however, reproducible performance is important and pulses that are longer than 1000 fs are not good because they can induce thermal damage. The entry of the pulse shaper into this application can let companies work with less expensive laser platforms that rely on the pulse shaper to ensure perfect laser performance. The pulse shaper can monitor deviations, and in fact, the pulse shaper can diagnose the laser well before it stops working. The move toward less expensive lasers will also allow the development of refractive surgical units that have two or more femtosecond laser modules. If at any point one of the lasers stops working, the system automatically switches to the second module. The computer can then alert the company to send a replacement module and document the type of failure. It is possible, but has yet to be proven, that shaped femtosecond pulses can optimize the different procedures performed by the refractive surgical unit. The two main functions are to cut the corneal flap and to ablate the cornea to correct its refractive properties. The pulse shaper could generate multiple pulses such that one intense pulse is broken into several pulses in order to achieve faster gentler ablation. Another possibility would be to create a pulse with a fast rise to seed ions in the tissue and then a long picosecond plateau that drives the electrons and causes cascaded ionization and ablation. The optimized femto-LASIK system will be introduced once the companies that came early into this market have recovered their early investment and are looking forward to distinguishing themselves from others.

One criticism I hear from ophthalmologists about the early units is that they consider femtosecond laser-based surgical systems to be very expensive. They would be much more willing to make the investment if these systems were capable of performing a number of different surgical procedures. The eye presents multiple challenges that could be addressed by a versatile femtosecond ophthalmology system. For example, the unit could perform refractive correction, cataract (lens) removal, and retinal procedures such as macular degeneration treatments. For refractive correction, the laser needs to work at very shallow depths, of about a tenth of a millimeter into the eye. For cataracts, the laser needs to deliver energetic pulses to break down the lens of the eye while avoiding collateral damage. For retinal procedures, the laser needs to arrive at the cornea and stop the progression of macular degeneration with microscopic three-dimensional precision. This latter application needs to take into account the potential temporal broadening effects of the dispersion that comes from transmitting through the entire eye. The pulse shaper can be used to achieve the level of versatility required for such a multifunction ophthalmic surgical unit.

Continuing on the surgical theme, one can consider other types of surgeries that could benefit from femtosecond lasers. Let's first remember that femtosecond laser pulses have high peak intensities but low energies. That makes them ideal for very fine sub-millimeter control, but inefficient for removal of large structures. When thinking of fine structures, one can consider surgery of fine organs such as those involved in hearing, or removal of polyps in the vocal cords. In those cases, sub-millimeter accuracy is extremely important. It is possible that some aspects of brain surgery may one day involve the combination of multimodal imaging and femtosecond ablation. One femtosecond laser that is optimized by the pulse shaper could achieve both of these functions.

So far, we have considered instruments that could be quite compact and relatively inexpensive. We now consider systems that could take a substantially large room in the basement of a hospital. Here, I explore the generation of high-energy radiation using femtosecond laser pulses. The generation could be in the form of x-rays, gamma rays, high-energy electrons, high-energy protons, neutrons, and ions. While the femtosecond laser required to achieve this high-energy radiation is quite large, the technology has been proven already in laboratories. The investment to build such femtosecond laser exceeds a few million dollars; however, the benefits of such a system are quite numerous. In a hospital, such a system could be used for x-ray imaging, x-ray radiation treatments, gamma ray cancer treatments, and electron, proton, positron, ion, and neutron bombardment for cancer treatments. At the heart of the system could be a set of very reliable pump sources, a robust femtosecond laser, and a pulse shaper that keeps the output pulses continuously monitored and optimized. High-power amplification would be required to generate high-flux and high-energy photons. In addition, the system would use highly efficient particle acceleration to reach mega–electron volt energies required to penetrate deep into tissue to destroy cancerous tumors. Turns out that intense femtosecond pulses can be used to achieve highly efficient and compact particle acceleration. The process responsible for this is called wake-field acceleration, and it takes advantage of the fact that charged particles interact with strong electric fields within the pulse. Therefore, in the wake of an intense field, it is possible to accelerate particles to within fractions of the speed of light. Mega–electron volt particles have been found to be particularly useful for targeting and destroying cancerous tumors that are essentially inoperable. The type of system being envisioned for the creation of x-ray and gamma-ray photons could be used for difficult to reach cancerous tumors such as those in the brain. Interestingly, other non-medical applications have been proposed for these sources such as for monitoring the contents of shipping containers, in order to explore their contents and search for radioactive materials. Gamma rays can penetrate shipping containers and a table-top gamma ray source would greatly increase the number of containers whose contents are examined before entering a country.

Femtosecond laser technology was developed at Bell Laboratories motivated by the need for short pulses to enable fast communications. At the time this book was written, the bulk of fiber communication takes place with nanosecond laser pulses. The transition to picosecond pulses is already taking place. The need for speed, especially at locations where large amounts of data are aggregated and manipulated, is now reaching the point where transition to femtosecond pulses is required. The practical limits to entry of the technology are dispersion and nonlinear optical signal degradation. The pulse shaper can mitigate both of these processes. Only time will tell if pulse shapers will be widely implemented in femtosecond optical communications, or if the technological difficulties will be addressed through the use of specialty fibers such as hollow-core fibers that introduce much less dispersion and nonlinear optical signal degradation and digital signal processors capable of numerical dispersion compensation. Fast communications will also be required within computer networks at short (a few meters) or long (thousands of kilometers) distances. Specialty fibers and integrated optical laser systems where the pulse laser and pulse shaper are implemented on a computer chip will be used for short distances. For long distances, ground-to-satellite and satellite-to-ground communication can be carried out in free space with much less dispersion than on optical fibers.

So far, we have considered here industrial and medical large-scale implementations of pulse shaping. Scientific research into pulse shaping is still in its infancy. A number of directions are being considered including shaping of nonstatistical light. Non-statistical light is an expression used to describe light made of entangled photons. Imagine that a given photon is coherently decomposed into two photons with half the energy. The resulting two photons are said to be entangled, and their probability for re-forming the original higher-energy photon is much greater than the corresponding probability for two nonentangled photons. The optical process described for creating entangled photons is known as parametric down-conversion (PDC). Under the right optical conditions, PDC can generate broadband entangled pairs of photons with many similarities to femtosecond pulses. For example, such a source is susceptible to dispersion because of its broadband nature. Therefore, many of the concepts used here for spectral phase measurement and dispersion compensation are applicable to broadband sources of entangled photons, and their implementation is best carried out using pulse shapers.

Research is also being carried out on pulse shaping of incoherent light. Pulse shapers are very much like common path interferometers. Therefore, pulse shapers can be used whenever interference is part of a measurement. For example, there are multiple methods for interferometric spectroscopy (also known as Fourier Transform Spectroscopy). Many of those systems have included optical delay lines with mechanical actuators moving physical mirrors. Pulse shapers could provide similar functions without moving parts and with much greater accuracy.

The future of shaped ultrafast pulses will eventually move into integrated optical circuits (IOCs). These IOCs will include a femtosecond oscillator that, because of its size and stability, will function as a frequency comb. The IOCs will likely include a pulse shaper within the chip. Prototypes of these systems have already been built. As these systems are commoditized, as discussed in the next chapter, their price and widespread use will increase.

17 The Ultrafast Laser Scaling Revolution

The first femtosecond laser system I built (shown in Figure 0.1), starting in 1986, was a copy of the system built at Bell Laboratories in 1984. The system occupied an optical bench that was 20 × 5 ft and had eight different laser dyes circulating in various stages of amplification and saturable absorption. On a good day, it worked for several hours, but more typically, it stopped working at least once per hour. Thirty years later, there are femtosecond oscillators that are the size of a can of sardines that can withstand impacts associated with being dropped from the optical bench, and the systems keep working! Amplified laser systems with 10-mJ outputs now come in a box that measures 2 × 6 ft. These systems can function for months without supervision. This chapter deals with a class of lasers that has defied scaling. Even today, most commercial femtosecond laser sources are partially "handcrafted" or at least "hand tuned." This facts makes femtosecond lasers expensive and differentiates the lasers from each company by their design and skill at tuning the performance of their laser. However, the transition to scalable femtosecond lasers is now fully underway. Here, we discuss this transition and how, in analogy to electronic integrated circuits, photonic integrated circuits (PICs) will include a shaped femtosecond laser. As with electronics, PICs will lead to a dramatic drop in price for these devices.

Already between 2007 and 2017, the entry of femtosecond lasers to both ophthalmic clinics and to industrial micromachining has led to significant drops in price. At the time of this writing, an ophthalmic femtosecond fiber laser (just the laser) can be purchased for about $100,000. The market for thousands of these units should eventually lead to what is known in economics as commoditization. When a product is commoditized, price drives its sales, because the market has already set standards of operation. Examples of commoditization are cell phones, in which, unless one is looking for the absolute latest technology, the price is quite affordable. What makes cell phones so affordable considering the computational, communications, and sensing capabilities that are built in, is the combination of a market size close to a billion people and cell-phone signal access. In addition, cell phones are scalable. Scalability implies that a system can be manufactured in greater quantities without needing to scale human resources. The best example of scalability is the integrated circuit central processing unit (IC-CPU). First introduced in 1969, CPUs have grown exponentially in the number of transistors they include according to Moore's Law (doubling performance every 2 years or decreasing in price by 50% every 2 years), but we don't require billions of people to manufacture them.

Having introduced economic concepts such as commoditization and scalability, it is clear that diode lasers have seen their prices drop to a few cents per unit. However, femtosecond lasers have not followed Moore's Law. As indicated in Chapter 13, industrial applications are pushing prices down, but very modestly. I predict that femtosecond lasers will soon reach scalability and commoditization. Once achieved, the price and performance of femtosecond lasers will follow the accelerated rate of improvement similar to that which we see in electronics and communications systems.

How will the commoditized femtosecond laser look? While there are many potential designs already in use, I believe that communication technology will lead the way in the design of commoditized femtosecond lasers. Already, communication systems use commoditized 10-ps and 1-ps pulsed lasers. Designs for 100-fs lasers are already being tested. As discussed in Chapters 4 through 6, as the bandwidth gets broader, dispersion and phase control become critical. Therefore, the commoditized femtosecond laser I am envisioning will look like a modern PIC. In Chapter 14, we talked about a number of different multiplexing approaches that are already in use for optical communications. Most of those multiplexers already come as PICs. For example, the diagram in Figure 17.1 shows a 40-channel transmitter PIC, consisting of an array of 40 Gbps distributed feedback lasers with electro-absorption modulator integrated lasers, which are combined using an arrayed waveguide grating demultiplexer and high-speed photodiodes operating at 40 Gbps. Notice how similar the transmitter is to a femtosecond laser with a pulse shaper. These types of PIC devices have dimensions of approximately $10 \times 20 \times 5$ mm, and are quickly commoditized thanks to communication standards. They are known as either transmitter or receiver optical subassemblies (TOSA or ROSA).

Following the previous examples, one can conceive of a femtosecond laser PIC, such as that shown schematically in Figure 17.1. It includes a distributed Bragg

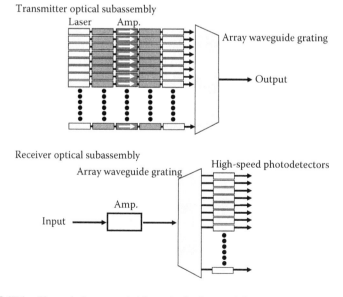

FIGURE 17.1   Photonic integrated chip optical subassemblies.

resonator laser, a modulator, a first amplifier, a micro-resonator, a second amplifier, an arrayed waveguide grating demultiplexer, an array of fast electro-optic modulators, an arrayed waveguide grating multiplexer, and a fiber output (Figure 17.2). While there are hundreds of specifics that need to be figured out, a system such as that being proposed here would be scalable and would achieve commoditization by conforming to a number of communication standards. For example, among the required parameters would be 100-fs pulse duration and 1024 independently phase-modulated wavelengths (also known as a frequency comb) within the spectral bandwidth of the pulse. Such a system is scalable because it can be manufactured entirely by machines and because the output is fully controllable through the externally addressable modulators without movable parts. According to what we have discussed in terms of dispersion, pulse characterization, and automated pulse compression in Chapters 4, 5, 8, 9, and 10, this laser would be capable of maintaining optimum performance without moving parts or external intervention. If adopted for communications, its price would quickly drop from thousands to a few hundreds of dollars because companies could stamp hundreds or thousands of these units per day. The stability and mean time between failures of such a system would be from tens to hundreds of thousands of hours, well beyond what we have ever imagined for femtosecond lasers. A communication server could include multiple such femtosecond lasers and depend on them continuously for communication at terabits per second.

As soon as the femtosecond laser chip is commoditized, it is to be expected that companies will offer compatible pre-amplification and power amplification modules that will allow integrator companies to configure the shaped femtosecond lasers of the future (Figure 17.2). Amplification will likely involve some form of chirped-pulse amplification, that patent would have expired, so this can

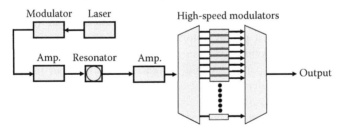

Femtosecond laser optical subassembly (FLOSA)

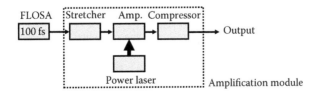

Modular femtosecond laser concept

FIGURE 17.2 Photonic integrated chip femtosecond laser and modules.

be accomplished without great expense. The power will come from diode lasers or diode-pumped fiber lasers. The price for the amplifier lasers is now near $1 per watt, and it will continue to drop by 50% every 2 years. The generation of very short pulses can be accomplished by using highly nonlinear fiber to create broader bandwidth through self-phase modulation (Chapter 6). When amplified ultrashort (5–10 fs) pulses are desired, then optical parametric amplification can be used to amplify the pulses while maintaining the bandwidth. Pulse compression of the final output will be accomplished by a pair of gratings, or by the use of second-generation phase compression optics being developed at present. The modern femtosecond laser systems will take advantage of the adaptive pulse shaper in the femtosecond laser chip for optimization of the final output, even after amplification. In fact, it is very likely that the femtosecond laser chips used for communication will make it easy to achieve carrier-envelope-phase stabilization of the output thanks to their micro-resonator and the fast-phase modulators on-board.

The applications discussed in this book (Chapters 11 through 16) are all very sensitive to price. With scientific femtosecond lasers presently at $100,000 and up to millions of US dollars, the cost of the laser is typically a limiting factor. The development of ultrastable femtosecond laser oscillators with pulse-shaping capabilities that cost $1000, and amplification modules that scale at $1000 per watt of continuous power, will greatly accelerate the femtosecond laser applications discussed in this book and many others that have yet to be invented. While I am offering here one vision for how femtosecond lasers may become commoditized, this is not the only path. There are some exciting developments occurring in a class of laser called "gain-switched" diodes. For these, the diode laser power is switched quickly so that sub-picosecond pulsed emission is achieved. These sources do not provide the wavelength stability, precision, coherence, and polarization required for communications or for most scientific applications. However, gain-switched diodes may provide an inexpensive seed for industrial lasers operating at pulse durations greater than 100 fs. I would imagine that gain-switched diodes will have their period of time when they are used for industrial applications; however the communications femtosecond PICs will be much more stable, flexible, and less expensive within 10 years. Like all of our information-based technologies (computers, cell phones), the femtosecond PICs will be configurable and upgradeable by software.

The overall message of this book is that femtosecond laser systems equipped with a programmable pulse shaper provide the most advanced energy source because it can be configured to emit pulses of any duration (down to attoseconds and zeptoseconds in the near future) and any wavelength (from microwaves to x-rays and gamma rays in the future). When amplified femtosecond lasers were first developed, with models very similar to the one shown in Figure 0.1, it was very difficult to envision applications. I hope the reader can now appreciate the role that pulse shaping has played in femtosecond laser pulse characterization, compression, and optimization, as well as how pulse shaping is critically important for enabling the commoditization of the femtosecond laser, and finally, how pulse shaping can improve most femtosecond laser applications in a number of areas including medical diagnosis and therapy, communications, and industrial material processing. The pulse shaper will enable autonomous systems where an artificial intelligence system controls the laser to achieve certain desired goals. We can envision how, through laser micro-processing, femtosecond lasers will be used to create ever-more advanced shaped femtosecond laser photonic integrated chips and other femtosecond laser modules.

Further Reading

REVIEWS/BOOKS ON PULSE SHAPING

Y. Coello, V. V. Lozovoy, T. C. Gunaratne, B. W. Xu, I. Borukhovich, C. H. Tseng, T. Weinacht, and M. Dantus, "Interference without an interferometer: a different approach to measuring, compressing, and shaping ultrashort laser pulses," *J. Optic. Soc. Am. B Optic. Phys.* **25**, A140–A150 (2008).

D. Goswami, "Optical pulse shaping approaches to coherent control," *Phys. Rep. -Rev. Sec. Phys. Lett.* **374**, 385–481 (2003).

V. V. Lozovoy and M. Dantus, "Coherent control in femtochemistry," *Chem. Phys. Chem.* **6**, 1970–2000 (2005).

A. M. Weiner, "Femtosecond pulse shaping using spatial light modulators," *Rev. Sci. Instrum.* **71**, 1929–1960 (2000).

A. M. Weiner, *Ultrafast Optics*, Wiley Series in Pure and Applied Optics (John Wiley & Sons, Inc., Hoboken, NJ, 2009).

SELECTED PAPERS ON SHAPER BASED PULSE CHARACTERIZATION AND COMPRESSION

T. Baumert, T. Brixner, V. Seyfried, M. Strehle, and G. Gerber, "Femtosecond pulse shaping by an evolutionary algorithm with feedback," *Appl. Phys. B Lasers Opt.* **65**, 779 (1997).

Y. Coello, V. V. Lozovoy, T. C. Gunaratne, B. Xu, I. Borukhovich, C. -h. Tseng, T. Weinacht, and M. Dantus, "Interference without an interferometer: a different approach to measuring, compressing, and shaping ultrashort laser pulses," *J. Opt. Soc. Am. B* **25**, A140-A150 (2008).

A. Galler and T. Feurer, "Pulse shaper assisted short laser pulse characterization," *Appl. Phys. B Lasers Opt.* **90**, 427 (2008).

V. Loriot, G. Gitzinger, and N. Forget, "Self-referenced characterization of femtosecond laser pulses by chirp scan," *Opt. Express* **21**, 24879–24893 (2013).

V. V. Lozovoy, G. Rasskazov, D. Pestov, and M. Dantus, "Quantifying noise in ultrafast laser sources and its effect on nonlinear applications," *Opt. Express* **23**, 12037–12044 (2015).

M. Miranda et al., "Characterization of broadband few-cycle laser pulses with the d-scan technique," *Opt Express* **20**, 18732–18743 (2012).

D. Pestov, V. V. Lozovoy, and M. Dantus, "Single-beam shaper based pulse characterization and compression using MIIPS sonogram," *Opt. Lett.* **35**, 1422–1424 (2010).

G. Rasskazov, V. V. Lozovoy, and M. Dantus, "Spectral amplitude and phase noise characterization of titanium-sapphire lasers," *Opt. Express* **23**, 23597–23602 (2015).

P. Schlup and R. A. Bartels, "Impact of measurement noise in tomographic ultrafast retrieval of transverse light E-Fields (TURTLE) ultrashort polarization characterization," *IEEE Photon. J.* **1**, 163 (2010).

B. von Vacano, T. Buckup, and M. Motzkus, "In situ broadband pulse compression for multiphoton microscopy using a shaper-assisted collinear SPIDER," *Opt. Lett.* **31**, 1154–1156 (2006).

D. E. Wikcox and J. P. Ogilvie, "Comparison of pulse compression methods using only a pulse shaper," *J. Opt. Soc. Am. B* **31**, 1544–1554 (2014).

B. W. Xu, J. M. Gunn, J. M. Dela Cruz, V. V. Lozovoy, and M. Dantus, "Quantitative investigation of the multiphoton intrapulse interference phase scan method for simultaneous phase measurement and compensation of femtosecond laser pulses," *J. Opt. Soc. Am. B Opt. Phys.* **23**, 750 (2006).

T. Wu, J. Tang, B. Hajj, and M. Cui, "Phase resolved interferometric spectral modulation (PRISM) for ultrafast pulse measurement and compression," *Opt. Express* **19**, 12961–12968 (2011).

PAPERS ON PULSE SHAPING TECHNIQUES

M. Dantus and K. Monro, "Ultrafast temporal shaping is coming of age," *Biophotonics* **21**, 24–28 (2014).

J. Extermann, S.M. Weber, D. Kiselev, L. Bonacina, S. Lani, F. Jutzi, W. Noell, N.F. de Rooij, and J.-P. Wolf, "Spectral phase, amplitude, and spatial modulation from ultraviolet to infrared with a reflective MEMS pulse shaper," *Opt. Express* **19**, 7580 (2011).

F. Ferdous et al., "Spectral line-by-line pulse shaping of on-chip microresonator frequency combs", *Nat. Photon.* **5**, 770 (2011).

E. Frumker and Y. Silberberg, "Phase and amplitude pulse shaping with two-dimensional phase-only spatial light modulators," *J. Opt. Soc. Am. B* **24**, 2940–2947 (2007).

A. Konar, V.V. Lozovoy, and M. Dantus, "Solvent environment revealed by positively chirped pulses," *J. Phys. Chem. Lett.* **5**, 924–928 (2014).

V.V. Lozovoy, G. Rasskazov, A. Ryabtsev, and M. Dantus, "Phase-only synthesis of ultrafast stretched square pulses," *Opt. Express* **23**, 27105–27112 (2015).

D.S. Moore, S.D. McGrane, M.T. Greenfield, R.J. Scharff, R.E. Chalmers, "Use of the Gerchberg–Saxton algorithm in optimal coherent anti-Stokes Raman spectroscopy," *Anal. Bioanal. Chem.* **402**, 423 (2012).

D. Pestov, V. V. Lozovoy, and M. Dantus, "Multiple independent comb shaping (MICS): Phase-only generation of optical pulse sequences," *Opt. Express* **17**, 14351 (2009).

D. Pestov, A. Ryabtsev, G. Rasskazov, V.V. Lozovoy, and M. Dantus, "Real-time single-shot measurement and correction of pulse phase and amplitude for ultrafast lasers," *Opt. Eng.* **53**, 051511 (2014).

G. Rasskazov, A. Ryabtsev, V.V. Lozovoy and M. Dantus, "Laser-induced dispersion control," *Opt. Lett.* 39 (2014).

I. Saytashev, B. Xu, M.T. Bremer, and M. Dantus, "Simultaneous selective two-photon microscopy using MHz rate pulse shaping and quadrature detection of the time-multiplexed signal," *Ultrafast Phenomena XIX*, K. Yamanouchi et al., Eds. (Springer Proceedings in Physics 162, 2015).

J.W. Wilson, P. Schlup, and R.A. Bartels, "Ultrafast phase and amplitude pulse shaping with a single, one-dimensional, high-resolution phase mask," *Opt. Express* **15**, 8979–8987 (2007).

OTHER INTERESTING REFERENCES

N. Accanto, J.B. Nieder, L. Piatkowski, M. Castro-Lopez, F. Pastorelli, D. Brinks, N.F. van Hulst, "Phase control of femtosecond pulses on the nanoscale using second harmonic nanoparticles," *Light Sci. Appl.* **3**, e143 (2014).

N. Accanto, L. Piatkowski, I. M. Hancu, J. Renger, N. E. van Hulst, "Resonant plasmonic nanoparticles for multicolor second harmonic imaging," *Appl. Phys. Lett.* **108**, 083115 (2016).

M. Balu, I. Saytashev, J. Hou, M. Dantus, and B.J. Tromberg, "Sub-40 fs, 1060-nm Yb-fiber laser enhances penetration depth in nonlinear optical microscopy of human skin," *J. Biomed. Opt.* **20**, 120501 (2015).

S. Berweger, J.M. Atkin, X.J.G. Xu, R.L. Olrnon, M.B. Raschke, "Femtosecond nanofocusing with full optical waveform control," *Nano Lett.* **11**, 4309 (2011).

M. Bremer, P. Wrzesinski, N. Butcher, V. V. Lozovoy, and M. Dantus, "Highly selective standoff detection and imaging of trace chemicals in a complex background using single-beam coherent anti-Stokes Raman scattering," *Appl. Phys. Lett.* **99**, 101109 (2011).

M.T. Bremer and M. Dantus, "Standoff explosives trace detection and imaging by selective stimulated Raman scattering," *Appl. Phys. Lett.* 103, 061119 (2013).

D. Brinks et al., "Visualizing and controlling vibrational wave packets of single molecules", *Nature* **466**, 905 (2010).

P. Devi, V.V. Lozovoy, and M. Dantus, "Measurement of group velocity dispersion of solvents using 2-cycle femtosecond pulses: Experiment and theory," *AIP Adv.* **1**, 032166 (2011).

H. Frostig, O. Katz, A. Natan, and Y. Silberberg, "Single-pulse stimulated Raman scattering spectroscopy," *Opt. Lett.* **36**, 1248 (2011).

A. Gamouras, R. Mathew, and K.C. Hall, "Optically engineered ultrafast pulses for controlled rotations of exciton qubits in semiconductor quantum dots," *J. Appl. Phys.* **112**, 014313 (2012).

L. Jiang, P. Liu, X. Yan, Ni Leng, Ch. Xu, H. Xiao, and Y. Lu, "High-throughput rear surface drilling of microchannels in glass based on electron dynamics control using femtosecond pulse trains," *Opt. Lett.* **37**, 2781 (2012).

O. Katz, E. Small, Y. Bromberg, and Y. Silberberg, "Controlling ultrashort pulses in scattering media," *Nat. Photon.* **5**, 372 (2011).

J. Kohler, M. Wollenhaupt, T. Bayer, C. Sarpe, T. Baumert, "Zeptosecond precision pulse shaping," *Opt. Express* **19**, 11638 (2011).

A. Konar, V.V. Lozovoy, and M. Dantus, "Solvation Stokes-shift dynamics studied by chirped femtosecond laser pulses," *J. Phys. Chem. Lett.* **3**, 2458–2464 (2012).

A. Konar, V.V. Lozovoy, and M. Dantus, "Solvent environment revealed by positively chirped pulses," *Ultrafast Phenomena XIX*, K. Yamanouchi et al., Eds. (Springer Proceedings in Physics 162, 2015).

A. Konar, J. Shah, V.V. Lozovoy, and M. Dantus, "Optical response of fluorescent molecules studied by synthetic femtosecond laser pulses," *J. Phys. Chem. Lett.* **3**, 1329–1335 (2012).

V.V. Lozovoy, G. Rasskazov, A. Ryabtsev, and M. Dantus, "Phase-only synthesis of ultrafast stretched square pulses," *Opt. Express* 23, 27105–27112 (2015).

R. Mittal, R. Glenn, I. Saytashev, V.V. Lozovoy, and M. Dantus, "Femtosecond nanoplasmonic dephasing of individual silver nanoparticles and small clusters," *J. Phys. Chem. Lett.* **6**, 1638–1644 (2015).

Y. Nabekawa et al., "Multi-terawatt laser system generating 12-fs pulses at 100 Hz repetition rate," *Appl. Phys. B* **101**, 523 (2010).

G. Rasskazov, A. Ryabtsev, V.V. Lozovoy, and M. Dantus, "Mitigating self-action processes with chirp or binary phase shaping," *Opt. Lett.* **41**, 131–134 (2016).

A.P. Rudhall et al., "Exploring the ultrashort pulse laser parameter space for membrane permeabilisation in mammalian cells," *Sci. Rep.* **2**, 858 (2012).

I. Saytashev, R. Glenn, G.A. Murashova, S. Osseiran, D. Spence, C.L. Evans, and M. Dantus, "Multiphoton excited hemoglobin fluorescence and third harmonic generation for non-invasive microscopy of stored blood," *Biomed. Opt. Exp.* **7**, 3449–3460 (2016).

H. Tu, Y. Liu, D. Turchinovich, and S.A. Boppart, "Compression of fiber supercontinuum pulses to the Fourier-limit in a high-numerical-aperture focus," *Opt. Lett.* **36**, 2315 (2011).

H. Tu, Y. Liu, D. Turchinovich, M. Marjanovic, J.K. Lyngsø, J. Lægsgaard, E.J. Chaney, Y. Zhao, S. You, W.L. Wilson, B. Xu, M. Dantus, and S. A. Boppart, "Stain-free histopathology by programmable supercontinuum pulses," *Nat. Photon.* **10**, 534–540 (2016).

P. Wrzesinski et al., "Binary phase shaping for selective single-beam CARS spectroscopy and imaging of gas-phase molecules," *J. Raman Spectrosc.* **42**, 393–398 (2011).

P. Wrzesinski et al., "Group-velocity-dispersion measurements of atmospheric and combustion-related gases using an ultrabroadband-laser source," *Opt. Express* **19**, 5163–5170 (2011).

H.J. Wu, Y. Nichyama, T. Narushima, K. Imura, and H. Okamoto, "Sub-20-fs time-resolved measurements in an apertured near-field optical microscope combined with a pulse-shaping technique," *Appl. Phys. Exp.* **5**, 062002 (2012).

Index

Page numbers followed by f and t indicate figures and tables, respectively.

T - #0149 - 111024 - C120 - 229/152/6 - PB - 9780367877446 - Gloss Lamination